INTEGRATION AND CONTROL
OF METABOLISM

INTEGRATION AND CONTROL OF METABOLISM

Dr. Naa A. Adamafio, B.Sc. (Ghana), Ph.D. (Monash)
Senior Lecturer
Department of Biochemistry
University of Ghana
Ghana

Dr. Laud K. N. Okine, B.Sc. (Ghana), Ph.D. (Surrey)
Senior Lecturer
Department of Biochemistry
University of Ghana
Ghana

Dr. Jonathan P. Adjimani, B.Sc. (KNUST), M.Sc. (Brock), Ph.D. (Utah State)
Senior Lecturer
Department of Biochemistry
University of Ghana
Ghana

iUniverse, Inc.
New York Lincoln Shanghai

INTEGRATION AND CONTROL OF METABOLISM

iUniverse books may be ordered through booksellers or by contacting:

iUniverse
2021 Pine Lake Road, Suite 100
Lincoln, NE 68512
www.iuniverse.com
1-800-Authors (1-800-288-4677)

FIRST EDITION

ISBN-13: 978-0-595-34067-5 (pbk)
ISBN-13: 978-0-595-78852-1 (ebk)
ISBN-10: 0-595-34067-9 (pbk)
ISBN-10: 0-595-78852-1 (ebk)

Printed in the United States of America

PREFACE

The writing of this book has been prompted by the fact that as teachers of Intermediary Metabolism, Biochemistry of Hormones and Integration and Control of Metabolism we have not come across a book that deals specifically with the Integration and Control of Metabolism. Many students have also complained about the lack of a textbook that deals with the subject head on instead of it being scattered across various chapters in some textbooks while others treat it superficially.

This book is, therefore, aimed at bringing relief to undergraduate students in biochemistry and medical students in their pre-clinical years who want a concise and insightful textbook on Integration and Control of Metabolism to gain understanding as to how most cellular organisms integrate and control their metabolism. It is divided into three parts: 1) Fundamentals of Metabolic Control, 2) Control of Cellular Metabolism and 3) Integration of Metabolism.

Part one is made up of two chapters. Chapter one deals with the concepts of metabolic control while chapter two deals with the role of regulatory enzymes in metabolic control. Living organisms are open systems with metabolic pathways, which comprise equilibrium and non-equilibrium reactions. Because of the presence of different energy fuels in different tissues, and different energy requirements at particular times (nutritional and physiological states), there is a need for metabolic regulation. Metabolic pathways are generally under hormonal and/or nervous control with the flow of fuels from one organ to the other so that the energy needs of individual organs are met.

In order to understand metabolic regulation, it is important to appreciate the roles played by enzymes in metabolic pathways particularly those enzymes involved in the control of the key metabolic steps (regulatory enzymes). The characteristics and kinetic properties of these regulatory enzymes and the modes of regulation of their activities (fine and coarse controls) will be discussed in detail with specific examples where possible. Fine control involves the short-term regulation of enzyme activity including the actions of allosteric effectors and substrates or products as well as covalent modification involving mostly phosphorylation and dephosphorylation of enzymes. However, coarse control represents the long-term regulation of enzyme activity comprising the induction or repression of enzyme protein synthesis *de novo*.

The regulation of the metabolism of some major biomolecules, e.g. carbohydrates, lipids and nitrogen containing compounds is dealt with in part two, which comprises chapters three, four and five. The regulation of the synthesis and degradation of some of these biomolecules will be discussed in some detail to bring to the fore the key enzymes involved in the regulation of their metabolism and the factors that influence the kinetics of these enzymes as well as the modes of their regulation. However, in this section we shall treat regulation of some major metabolic pathways as if they are discrete units independent of each other.

In some cases, particularly with respect to nitrogen-containing compounds, regulation of the interconversions of compounds will also be discussed. The role of hormones and tissue differences in metabolic control will also be considered. As far as possible, regulatory steps in metabolic pathways of eukaryotic cells will be discussed. However, where other regulatory steps in prokaryotic metabolic pathways are of significance but not present in eukaryotic cells, these will be considered in order to reinforce our understanding of the roles played by regulatory enzymes in metabolic control.

In part two we have treated metabolic pathways as discrete metabolic units but there are some limitations to this.

- First, pathways do most often interact and even overlap for several reactions. The starting compounds and end products of a pathway are themselves products and precursors, respectively of other pathways.

- Second, the effects of regulatory molecules may link pathways in cells, thus blurring the division between one pathway and another.

- Finally, a sequence of reactions in one organ may operate simultaneously but in opposite direction in another organ. Intermediates flowing from one organ to the other complete an inter-organ cycle.

Metabolic routes can, therefore, be considered in the light of the overall metabolism of the organism, including the integration of multiple pathways and their regulation. Thus, part three deals with how the metabolic pathways within and between tissues/organs are integrated. One can gain insight into the interrelationships of the major metabolic pathways by looking at the changes in metabolism that occur in (a) a feed-starve cycle, (b) a running athlete and (c) some diseased states.

It is our hope that the first edition of this book will meet the aspirations of students and teachers alike in the subject area.

NAA/LKNO/JPA

ACKNOWLEDGEMENTS

The authors wish to acknowledge those who in diverse ways contributed towards the preparation and publication of this book.

First, our thanks go to Mrs. Christina Nettey and Mrs. Hetty Antwi-Bosiako of the Department of Biochemistry for typing the manuscript and to Mr. Adolf Kofi Awua (B.Sc.) also of the Department of Biochemistry for the drawing of the various illustrations presented in the book.

We are also thankful to all our colleagues in the Department of Biochemistry who through proofreading, corrections and valuable suggestions helped to improve this book.

Finally, our sincerest gratitude goes to our Publishers whose encouragement and support made the publication of this book possible.

CONTENTS

CHAPTER FOUR

LIPID METABOLISM 67

PART ONE
FUNDAMENTALS OF METABOLIC CONTROL

CHAPTER ONE

CONCEPTS OF METABOLIC CONTROL

Matter ingested serves as: a) a source of energy and b) a source of building materials. The energy can only be obtained after matter has been transformed through chemical reactions. These reactions are usually organized in sets referred to as metabolic pathways. The rate at which matter passes through a pathway is termed the flux through the pathway. Metabolic pathways exist as open systems.

OPEN AND CLOSED SYSTEMS

Closed System

A closed system is one that does not exchange matter or energy with the environment. Such a system will eventually reach equilibrium.

Open System

An open system is one in which there is a constant exchange of matter and energy with its environment. Such a system exists in all living organisms and is maintained continually in a non-equilibrium state.

Near-equilibrium and Non-equilibrium Reactions

Rapid changes in the rates of metabolic pathways are made possible by their basic design. Studies have shown that metabolic pathways consist of two types of reactions, namely near-equilibrium and non-equilibrium reactions. The majority of reactions in a given metabolic pathway are near-equilibrium. Non-equilibrium reactions are few, often positioned at the beginning of a metabolic pathway, and serve as regulatory points.

Short-term regulation of a pathway involves manipulation of the catalytic efficiency of a key enzyme that catalyses a non-equilibrium reaction. This increases the rate of reaction and the flux through the pathway because more substrates are made available for the near-equilibrium reactions in that pathway. Hormonal signals often lead to the manipulation of regulatory enzymes.

Near-equilibrium reactions

Equilibrium is reached when the rate of the forward reaction equals that of the reverse reaction, i.e. there is no net flux in either direction.

$$A + B \xrightleftharpoons[k_{-1}]{k_1} C + D$$

Rate of forward reaction $(R_1) = k_1$ [A] [B]
Rate of reverse reaction $(R_2) = k_{-1}$ [C] [D]
    At equilibrium:
Rate of forward reaction (R_1) = Rate of reverse reaction (R_2)

i.e. k_1 [A] [B] $= k_{-1}$ [C] [D]

$$\frac{k_1}{k_{-1}} = \frac{[C][D]}{[A][B]} = K_{eq} \text{ (equilibrium constant)}$$

The general features of near-equilibrium reactions are as follows:

- Catalysed by high capacity enzymes
- Substrate is converted to product as fast as it is supplied.
- Mass-action ratio (MAR) is, therefore, close to K_{eq}.
- Gibb's free energy change (ΔG) is usually small.
- Reaction is readily reversible, e.g. by a small increase in product concentration.
- There are many unoccupied active sites. Such reactions are, therefore, described as substrate-limited.
- Fine and coarse control measures that enhance enzyme activity do not influence reaction rate.
- Reaction rate increases when substrate concentration increases.

Non-equilibrium reactions

In general, metabolic processes in organisms are open systems in which there is a net flux and thus the overall process is removed from equilibrium. The direction of flux in the metabolic pathway is dictated by the state of the equilibrium at the regulatory steps.

A process that is far from equilibrium will undergo a negative free energy change (i.e. ΔG = -ve). Some of this free energy may be lost as heat to the environment so that the reaction proceeds until large proportions of the reactants are converted to products. Thus, for a pathway to proceed in one direction, there is need for the loss of some of the energy as heat. The balance can be converted to chemical energy by the cell as ATP. For example, in glycolysis the conversion of one mole of glucose to lactate is an 11-step reaction that ends in the production of two net moles of ATP.

The flux is not only governed by this degree of displacement from equilibrium (heat loss is driving force) but also dependent on other factors like: (a) activities of specific regulatory enzymes in the pathway or (b) the concentrations of cofactors in the pathway.

The general features of non-equilibrium reactions are as follows:

- Catalysed by low capacity enzymes
- Perpetual backlog of substrate
- MAR is far below K_{eq}.
- ΔG is usually large and negative.
- Reaction is essentially irreversible *in vivo*.
- The reaction rate can only be increased by enhancing catalytic efficiency or increasing enzyme quantity since the enzyme operates at V_{max}. Such reactions are described as enzyme-limited.
- Any increase in substrate concentration has no effect on rate of reaction since the enzyme is saturated.

Steady state

A condition in which substrate is continually supplied and products continually removed so that the concentrations of the intermediates remain constant despite a flux through the pathway is known as a steady state (ss). For example, glycolysis is maintained continually in a non-equilibrium steady state by a constant exchange of matter and energy between living organisms and their environment.

Whether an enzyme-catalysed reaction is near-equilibrium or in a state of non-equilibrium can be determined by comparing the established K_{eq} for the reaction with the MAR, as it exists within a cell. Studies have shown that the MAR of non-equilibrium reactions is 100 to 10,000 times lower than the K_{eq}.

In a reaction:

$$A + B \rightleftharpoons C + D$$

$$K_{eq} = \frac{[C][D]}{[A][B]}$$

the concentrations of the substances may change at any point in time but the K_{eq} remains a constant. The MAR is calculated in a similar manner except that the steady state (ss) concentrations of reactants and products within the cell are used in the equation:

$$MAR = \frac{[C]_{ss}[D]_{ss}}{[A]_{ss}[B]_{ss}}$$

NEED FOR REGULATION

In order to facilitate the efficient regulation of metabolic pathways, specific hormones are released under specific metabolic conditions. These hormones send signals to target cells and bring about appropriate changes in the flux through metabolic pathways. In addition, the intracellular concentrations of certain biomolecules affect the performance of key enzymes whose activities influence the overall rates of metabolic pathways.

Despite the presence of enzymes and metabolites in each cell, cellular metabolism is not random but is highly regulated. However, to appreciate the importance of individual metabolic pathways and their regulation within cells, these pathways must be viewed in the light of the whole organism.

For various reasons the flux through a pathway has to be controlled. These reasons include:

- The fact that organisms feed intermittently.
- The need to maintain homeostasis (e.g. fairly constant levels of essential metabolites).
- The need to satisfy the peculiar demands of various tissues and organs.
- Variations in the level of physical activity.

Intermittent Feeding

Most organisms feed at intervals. However, the requirements of cells for energy and building materials are fairly constant. Consequently, there are mechanisms that ensure that excess fuel molecules are stored after a meal and released when required. The rate of utilization of fuel molecules such as glucose is high in times of abundance and low in times of starvation. The ability of organisms to alter the rate of degradation of glucose and other fuel molecules depends on numerous factors. These include the release of insulin or glucagon from the pancreas and subsequent activation or inactivation of key enzymes in intermediary metabolism.

Homeostasis

One of the distinguishing characteristics of biological organisms is their ability to maintain a fairly constant internal environment. For example, blood glucose is maintained close to 5 mM. After a carbohydrate rich meal blood glucose concentration may rise to 12 mM and beyond. Under such circumstances virtually all cells increase their rate of glucose utilisation in order to bring down blood glucose levels. Conversely, in situations where blood glucose level falls because no glucose enters the bloodstream from the gut, the utilisation of glucose is minimized to conserve glucose for the brain. Furthermore, the liver sets in motion processes that result in the synthesis of glucose from non-carbohydrate sources. Clearly, the ability to maintain glucose homeostasis depends on mechanisms that alter the rate of glucose metabolism depending on metabolic circumstances.

Tissue/Organ Energy Needs

The existence of specialized cells in higher animals facilitates the division of labour. Brain, muscle, liver and adipose tissue differ not only in their functions but also in their energy requirements and fuel preferences. For example brain utilizes glucose, as the major source of energy while resting muscle prefers fatty acids. Since the composition of food ingested may vary from high carbohydrate to high fat or high protein, it may be necessary to alter the relative amounts of the various fuel molecules in order to satisfy the demands of different cell types. This is the responsibility of the liver and involves manipulation of the rates of various metabolic pathways in hepatocytes.

Physical Activity

Energy consumption is directly proportional to the intensity of physical activity. Obviously, because biological organisms do not undergo a constant level of physical activity the consumption of energy varies from time to time. The rates of energy-generating metabolic pathways are adjusted accordingly.

TISSUE ENERGY STORES

Energy reserves in mammals are stored in various tissues. These energy reserves are glycogen, triacylglycerols (TAGs) and proteins. Most of the glycogen and protein are stored in the muscle while most of the TAGs are stored in the adipose tissue (Table 1.1).

Table 1:1 Fuel reserves in a normal 70 kg human

Tissue	kcal (kJ) Fuel Reserve		
	Glycogen	Triacylglycerols	Mobilizable Protein
Adipose	80 (335)	135,000 (564,840)	40 (167)
Brain	8 (34)	0	0
Liver	400 (1,674)	450 (1,883)	400 (1,674)
Muscle	1,200 (5,021)	450 (1,883)	24,000 (100,416)

Source: Data from G.T. Cahill (1976) Clin. Endocrinol. Metab. 5: 59

Glycogen

Glycogen is a much more readily available energy source than TAGs. Thus, the use of TAGs is held in check until most of the glycogen stores in liver and muscle have been utilised. Generally, glycogen reserves in the liver are readily available to other tissues. This is because the liver possesses the enzyme glucose 6-phosphatase (G-6-Pase). This enzyme is capable of converting glucose 6-phosphate (G-6-P), an intermediary product of glycogenolysis, to glucose that can be released into the blood and transported to other peripheral tissues like brain that is in constant need of glucose.

Triacylglycerol

Once glycogen stores have been depleted from the liver and muscle, the TAG stores of the adipose tissue meet the energy needs of the body. The TAGs are first hydrolysed by lipases to fatty acids (FAs) and glycerol. The FAs can be utilised in the adipose tissue or transported to other tissues like the liver and muscle where they undergo β-oxidation to produce acetyl-CoA, which can enter the citric acid cycle (CAC) to generate energy for cellular processes.

Protein

The third major fuel reserve, protein, which comes mainly from the muscle can be hydrolysed to amino acids. The amino acids can be transported to the liver where they undergo oxidative deamination and transamination reactions to produce CAC intermediates for energy or glucose production. This use of protein stores causes wasting of the muscle and rarely occurs except in periods of prolonged starvation.

TISSUE FUNCTION

Each mammalian tissue has a specific function to perform depending on its anatomical and metabolic features. Each tissue makes characteristic demands and/or contributions to the energy pool. Some are mainly energy suppliers (e.g. adipose tissue). Others are mainly energy consumers (e.g. brain) and still others are both suppliers and consumers of energy (e.g. muscle and liver). Signals can be sent to energy-producing tissues about the nedds of energy-requiring tissues through chemical messengers, hormones, produced by endocrine glands. Hormones regulate the metabolic activities of different mammalian tissues, and coordinate all metabolic activities.

Adipose tissue

Adipose tissue maintains vast fuel reserves in the form of TAGs. It makes up 15% of the body mass of a young adult human with 65% being TAGs. Adipose tissue cells (adipocytes) are metabolically active and respond quickly to hormonal stimuli.

An adipocyte has an active glycolytic pathway, uses the CAC to oxidize acetyl-CoA from FAs and undertakes oxidative phosphorylation. During high carbohydrate dietary intake, adipose tissue converts glucose via pyruvate and acetyl-CoA into FAs from which TAGs are made and stored. In humans, however, FA synthesis occurs mainly in the liver. Adipocytes store TAGs arriving from the liver and also from the intestine after a fatty meal.

The TAGs stored in adipose tissue are hydrolysed by lipases to free fatty acids (FFAs) and glycerol, in a process called lipolysis, when fuel is needed. These are delivered to liver, skeletal and cardiac muscles via the bloodstream. Lipolysis is greatly enhanced by the hormone epinephrine, which activates hormone sensitive lipase (HSL). Glucagon exerts a similar effect both is less potent. Insulin has the opposite effect of epinephrine.

Brain

The brain makes no contributions to the fuel needs of an organism. The brain of resting humans uses about 50% of the glucose in supply by absorbing it from the bloodstream. It has an active respiratory metabolism and its oxygen requirements account for about 20% of the total oxygen consumed at rest. It metabolises glucose to CO_2 and H_2O by a combination of glycolysis, CAC and oxidative phosphorylation.

In times of starvation or stress, the brain's demand for glucose must be met first should the blood glucose fall below a critical level. In prolonged starvation the brain utilizes ketone bodies (β-hydroxybutyrate and acetoacetate), produced from FA oxidation, minimizing the use of muscle proteins.

Liver

The liver is the central organ responsible for all energy-related metabolism. One primary function of the liver is to maintain blood glucose at a reasonable level for absorption into most tissues. Thus, in a well-fed state, when glucose is abundant, the liver absorbs glucose using some for the provision of the energy and storing the excess as glycogen for future energy needs.

When blood glucose levels fall, such as in a fasted or starved state, the liver converts the glycogen to glucose (in glycogenolysis), which it secretes into the bloodstream to peripheral tissues that need it. When liver glycogen reserves are exhausted and blood glucose levels are low, the liver synthesizes glucose from lactate obtained from anaerobic glucose metabolism in erythrocytes and muscle, or amino acid precursors obtained from muscle proteolysis and glycerol from adipose tissue lipolysis . The liver also oxidizes FAs to acetyl-CoA at rates, which far exceed their utilization in the CAC, and, therefore, the acetyl-CoA units come together to form ketone bodies. The ketone bodies are carried to other tissues like the brain via the blood as sources of energy.

Muscle

Skeletal

Skeletal muscle operates under both anaerobic and aerobic conditions. The energy consumption and type of fuel used in the generation of energy (glucose, FAs or ketone bodies) depends on the intensity and duration of muscular activity.

In resting muscle the primary fuels are FAs obtained from the diet or from adipose tissue and ketone bodies from the liver. These are oxidized to acetyl-CoA, which enter the CAC for oxidation to CO_2. The ensuing transfer of electrons to O_2 in oxidative phosphorylation leads to the formation of H_2O and ATP. During intense activity, muscle operates anaerobically because the fast twitch muscle fibres involved have few mitochondria and the demand for ATP far exceeds the rate of O_2 delivery to these muscle cells. Anaerobic glycolysis is, therefore, the most important source of ATP to the muscle.

In less active muscles, the slow twitch muscle fibres characterised by abundant mitochondria and high levels of myoglobin operate aerobically. The sources of energy for the muscle include glucose produced by muscle glycogenolysis, which is metabolised to CO_2 and H_2O via glycolysis and the CAC. Aerobic skeletal muscles largely obtain energy from FAs, which supply as much as 60% of total energy needs for at least four hours of long distance running.

Cardiac

The heart muscle has a completely aerobic metabolism at all times. This organ, which abounds in mitochondria, can utilize a mixture of glucose, FAs and ketone bodies to generate ATP to meet its energy needs.

HORMONAL REGULATION OF FUEL METABOLISM

Hormones override the normal cellular controls. Each hormone is synthesized in an endocrine gland and secreted under special circumstances (e.g. under influence of nervous impulses from the brain). Once secreted, the hormone diffuses throughout the entire organism but triggers reactions only in those cells that carry specific receptors for the hormone.

The coordination of metabolism in separate organs of mammals is achieved by the neuroendocrine system. Individual cells in one tissue sense a change in the organism's circumstances and respond by secreting an extracellular chemical messenger that passes to another cell, where it binds to a receptor and triggers a change in the second cell.

In neural signalling, a chemical messenger such as acetylcholine may travel across the synaptic cleft to the next neuron in a network. Hormones on the other hand are carried via the blood to distinct organs and tissues where they interact with their target cells to generate biological responses.

Roles of Specific Hormones

The blood glucose level is kept near 5.0 mM by the combined actions of insulin, glucagon, epinephrine and cortisol on metabolic processes in many tissues.

Epinephrine

During stressful situations, neuronal signals from the brain trigger the release of epinephrine and norepinephrine from the adrenal medulla. Epinephrine acts primarily on muscle, adipose tissue and liver as follows:

- Activation of glycogen phosphorylase and inactivation of glycogen synthase by cAMP-dependent phosphorylation of the enzymes leading to the conversion of liver glycogen to glucose.
- Promotion of anaerobic metabolism of glycogen by skeletal muscle into lactate.
- Stimulation of fat mobilisation in adipose tissue through activation of HSL.
- Stimulation of glucagon secretion and inhibition of insulin secretion leading to mobilisation of fuels and inhibition of fuel storage.

Glucagon

When blood glucose level falls below 5.0 mM, such as in fasting or starvation, this triggers secretion of glucagon and decreases insulin release. Glucagon acts as follows:

- Stimulation of net breakdown of liver glycogen by activating glycogen phosphorylase and inactivating glycogen synthase.
- Inhibition of glycolysis in liver and stimulation of gluconeogenesis.
- Stimulation of fat mobilisation in adipose tissue by activating HSL. The liberated FAs are exported to the liver and other tissues sparing glucose use for the brain.

Insulin

The increased presence of insulin in the blood signals high blood glucose. It helps reduce blood glucose levels by acting as follows:

- Stimulation of glucose uptake by muscle and adipose tissues.
- Stimulation of glycolysis in liver, muscle and adipose tissue and the inhibition of gluconeogenesis in the liver.
- Activation of glycogen synthase and inactivation of glycogen phosphorylase leading to synthesis of glycogen and its storage in liver and muscle.
- Stimulation of storage of excess fuel (glucose or FAs) as TAGs in adipose tissue.

Cortisol

In stressful situations (e.g. anxiety, fear, pain, low blood glucose) there is release of cortisol from the adrenal cortex. Cortisol appears to act in concert with other hormones by changing the rates of synthesis of certain enzymes. Its effects on metabolism in target cells are as follows:

- Stimulation of the release of FAs from stored TAGs in adipose tissue, which are transported to other tissues for use as fuels. Glycerol produced is used for gluconeogenesis in liver.
- Stimulation of proteolysis of non-essential muscle proteins. The amino acids produced are exported to the liver to be used as gluconeogenic precursors.
- Stimulation of synthesis of PEP carboxykinase, a key enzyme in liver gluconeogenesis. The glucose produced is stored in the liver or exported to tissues that need it.

The net effect is to raise blood glucose back to its normal level and to store glycogen.

COMPARTMENTATION OF METABOLIC PATHWAYS

Metabolic pathways in living organisms do not take place in the same compartment. The efficiency of cellular function is greatly improved through separation of metabolic sequences into compartments. Compartmentation provides a means of regulating metabolic activity within the cell. The CAC, which is a central metabolic pathway, is located in the matrix of the mitochondria and the availability of substrates from the CAC depends on their transport across the mitochondrial membrane.

Controlling the rate at which substrates are supplied to a compartment is used in regulating a metabolic pathway. This regulatory function is due to the selective permeability of the membranes to various metabolites within the cell. The fate of metabolites depends on their cellular location and how their transport from one compartment to another is regulated. Thus, FAs are transported from the cytosol into the mitochondria only when there is the need for energy. Otherwise they are esterified in the cytosol and used, for example, in membrane biogenesis or exported to the adipose tissue and stored. Acetyl-CoA is produced in mitochondria and exported to the cytosol for the synthesis of FAs and cholesterol.

Most biochemical processes are irreversible; others such as glycolysis and gluconeogenesis are for the most part reversible. The regulation of reversible pathways is made possible partly by containing some aspects of the pathway in different cellular compartments. The locations of major metabolic processes in cells are shown in Table 1.3.

Table 1.3 Cellular locations of metabolic processes

Metabolic Process	Location
Glycolysis	Cytosol
Pentose phosphate pathway (PPP)	Cytosol
Citric acid cycle (CAC)	Mitochondria
Oxidative phosphorylation	Mitochondria
Gluconeogenesis	Cytosol /Mitochondria
Glycogenesis	Cytosol
Glycogenolysis	Cytosol
FA oxidation (β-oxidation)	Mitochondria
FA synthesis	Cytosol
Urea cycle	Cytosol /Mitochondria
DNA synthesis	Nucleus
Protein synthesis	Ribosomes

Further reading

Devlin, T.M. (1992). Textbook of Biochemistry with Clinical Correlations. Wiley—Liss Inc., New York

Fell, D. (1997). Understanding the Control of Metabolism. Portland Press, Miami

Moran, L.A., Scrimgeour, K.G., Horton, R.H., Ochs, R.S., and Rawn, J.D. (1994). Biochemistry. Neil Patterson Publishers/Prentice Hall Inc., Englewood Cliffs, New Jersey

Ochs, R.S., Hanson, R.W., and Hall, J. (eds). (1985). Metabolic Regulation. Elsevier Science Publishing Co. Inc., New York

Suarez, R.K., Staples, J.F., Lighton, J.R., and West, T.G. (1997). Relationships Between Enzymatic Flux Capacities and Metabolic Flux Rates: Nonequilibrium Reactions in Muscle Glycolysis. *Proc. Natl. Acad. Sci. U.S.A.* 94:7065-7069

Zubay, L.G. (1998). Biochemistry. Wm. C. Brown Publishers, Boston

CHAPTER TWO

ENZYMES AND METABOLIC REGULATION

Intracellular regulation involves the control of metabolism by enzymes and the factors that control these enzymes. Certain enzymes in metabolic pathways control the flux through the pathways in which they participate. These are generally responsible for the overall metabolic control within the cell and are called **regulatory enzymes**. This chapter will mainly dwell on these enzymes and their modes of regulation.

ENZYMES

Enzymes are biocatalysts formed in the cell either as simple or conjugated proteins. They are responsible for several coordinated reactions of living systems. In the cell, enzymes are found either bound to membrane structures or freely dissolved in various cellular compartments. Some enzymes are found in specific tissues. For example the liver and kidneys are the only organs endowed with G-6-Pase in the human body.

Some distinctive characteristics of enzymes include their substrate specificity, great catalytic power and the regulation of their activity to meet cellular needs. Enzymes catalyse reactions many orders of magnitude (10^3–10^7 times) faster than uncatalyzed reactions and show high specificity in respect of the substrates they act upon. Some may act on a group of closely related substrates while others act only on particular substrates. Other enzyme-catalysed reactions are control points in cellular metabolism. The regulation of metabolism includes the alteration of cellular concentrations of enzymes, substrates and enzyme inhibitors or activators and the modulation of the levels of activity of regulatory enzymes.

All chemical reactions have a potential energy barrier, the activation energy (E_a), which must be overcome before reactants can be converted to products. An enzyme helps to make a reaction thermodynamically feasible by providing alternative pathways that lower the E_a of a reaction (Fig. 2.1).

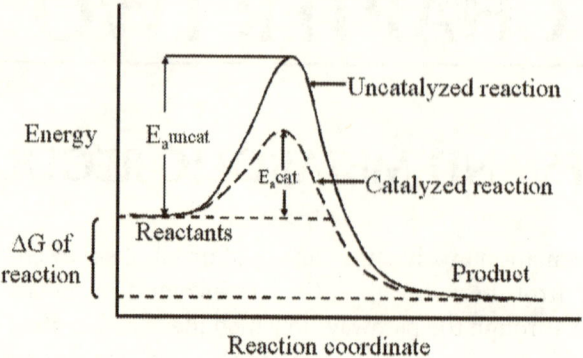

Fig. 2.1 Reaction coordinate diagram of catalysed and uncatalyzed reactions. Activation energy of catalysed reaction ($E_a$cat) is lower than that of the uncatalyzed reaction ($E_a$uncat).

An enzyme-catalysed reaction takes place within the confines of a pocket on the enzyme called the active site, where the substrate binds to be acted upon by the enzyme. In general, enzymes are much larger than their substrates. Only a small portion of substrate (a particular group, bond or linkage) interacts with substituent groups of some amino acid residues at the active site of the enzyme for the catalysis of its chemical transformation.

Enzymes that are involved in cellular regulation are usually oligomeric molecules that have separate binding sites for substrate and modulators, compounds that act as regulatory signals.

Properties of Enzymes

Most enzyme-catalysed reactions show saturable kinetics. Enzyme kinetics is the study of the rates of enzyme-catalysed reactions and provides indirect information regarding the specificities and catalytic mechanisms of enzymes. An enzyme (E) binds a substrate (S) to form an enzyme-substrate complex (ES) which then decomposes to product(s) (P) with the release of the enzyme. The kinetic variables for this simple enzymatic reaction can be written as:

$$E + S \underset{k_{-1}}{\overset{k_1}{\rightleftharpoons}} ES \xrightarrow{k_{cat}} E + P$$

The rate constants k_1 and k_{-1} give the rates of association of E with S and the dissociation of S from ES, respectively. The rate constant for the second step k_{cat} is the catalytic rate constant (turnover number) defined as the amount of substrate converted by one mole of an enzyme per unit time.

The ES can be converted to E and P. During the initial period of the reaction, [P] is low and, therefore, the conversion of E + P $\longrightarrow$ ES is negligible. The rate of reaction during this period is called the initial velocity (V). The key factor affecting the rate of an enzyme-catalysed reaction is the concentration of substrate [S]. For most enzymes, the relationship between V and [S] is a rectangular hyperbola (Fig. 2.2). This can be expressed mathematically by the Michaelis-Menten equation:

$$V = \frac{V_{max}\,[S]}{K_m + [S]}$$

It is an expression of the quantitative relationship between the rate of reaction (V), the maximum velocity (V_{max}), the initial substrate concentration ([S]) and the Michaelis constant (K_m).

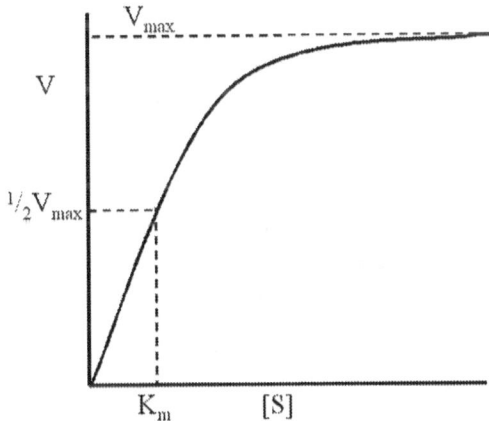

Fig. 2.2: Effect of substrate concentration on the rate of an enzyme-catalysed reaction, showing a normal hyperbolic curve. The K_m is the substrate concentration at ½ V_{max}.

Enzyme Inhibition

Enzymes are subject to inhibition. Inhibitors are compounds that bind to an enzyme leading to interference with its activity by preventing either the formation of the ES complex or its breakdown to E + P. Natural endogenous inhibitors serve as modulators of cellular metabolism whereas foreign enzyme inhibitors may function as drugs and many are used in the investigation of enzyme mechanisms.

There are two broad categories of enzyme inhibitors: (a) reversible and (b) irreversible inhibitors. The latter are bound covalently to enzymes rendering them inactive while the former (typical of endogenous inhibitors) are bound non-covalently to enzymes. Details of the modes of enzyme inhibition are beyond the scope of this book, except to say that most inhibitors of allosteric enzymes show a semblance of non-competitive inhibition.

REGULATORY ENZYMES

Regulatory enzymes respond to metabolic signals, by adjusting the flux of reactants through the metabolic pathways. A key regulatory enzyme is often located at an early step in a metabolic pathway and constitutes the committed step of the pathway serving as the point of regulation of the entire pathway. Inhibition of this step of a pathway plays an important role in the conservation of both material and energy.

Classes of Regulatory Enzymes

There are two major classes of regulatory enzymes in metabolic pathways:

a) Allosteric enzymes that function through reversible non-covalent binding of regulatory compounds called allosteric modulators or effectors, which are generally small metabolites or cofactors.

b) Enzymes that undergo reversible covalent modification such as phosphorylation and dephosphorylation.

Characteristics of Regulatory Enzymes

Regulatory enzymes have a number of general features:

- Their activities are sensitive to modulators/effectors (inhibitors or activators), which are often not structurally related to the substrates or products of the enzyme. These modulators bind non-covalently to the regulatory enzymes. They either modify the K_m or V_{max} of the enzymes they modulate but are themselves not changed chemically by the enzymes.

- Both classes of regulatory enzymes tend to be multi-subunit proteins with their regulatory site(s) and catalytic site(s) on separate subunits in some cases.

- They have at least one substrate for which a plot of V vs. [S] shows a sigmoidal curve rather than a hyperbolic one. A sigmoidal curve shows cooperativity in the binding of substrate, which indicates the presence of multiple substrate-binding sites in the enzyme.

- Most possess quaternary structure but not all enzymes with quaternary structure are regulatory.

CONTROL OF REGULATORY ENZYME ACTIVITY

The effectiveness with which individual regulatory enzymes in metabolic pathways operate must be controlled in a manner that reflects the availability of substrates, the utilisation of products and the overall needs of the cell. There are two major control systems in metabolic regulation involving enzymes: a) fine and b) coarse controls.

FINE CONTROL

Fine control involves short-term regulation of already synthesised key enzymes in metabolic pathways. This comprises three levels of regulation.

- Substrate/Product regulation
- Allosteric regulation involving cofactors, and metabolic intermediates.
- Covalent modification involving the reversible covalent binding of phosphoryl, acetyl, uridylyl groups to the hydroxyl groups of certain serine, threonine or tyrosine residues of the enzyme protein (see chapters 3 and 5).

Substrate/Product Control

This occurs through direct interaction of the substrate or end product with the enzyme at a regulatory step in a pathway. The higher the substrate concentration the more rapidly a reaction occurs. Conversely high levels of product, which can also bind to the enzyme, tend to inhibit the conversion of substrate to product. Thus, the product acts as a competitive inhibitor, e.g. the phosphorylation of glucose in glycolysis shown below.

$$\text{Glucose} + \text{ATP} \xrightarrow{\text{hexokinase}} \text{G-6-P} + \text{ADP}$$

The enzyme hexokinase, which catalyses the first step of glycolysis, is inhibited by G-6-P. If glycolysis is blocked for any reason, e.g. through the inhibition of phosphofructokinase-1 (PFK-1), G-6-P will accumulate leading to the inhibition of hexokinase and the reduction in the rate of entry of glucose into the pathway thus slowing down the entire glycolytic pathway. Substrate/product control of metabolic pathways may involve feedback or feed-forward controls of key enzymes in the pathway.

Feedback control

The cell can control the generation of the final product through activation or inhibition of an early step in the pathway. In a reaction:

$$A \longrightarrow B \longrightarrow C \longrightarrow D \longrightarrow E \longrightarrow F$$

it would be most efficient to slow down the first step so that A → B could be regulated by the amount of F. If the concentration of F is too high, it inhibits the conversion of A → B. The inhibition is lifted if the concentration of F decreases significantly. This process is called negative feedback control (feedback inhibition) and can be found in the conversion of threonine to isoleucine in which the first enzyme threonine dehydratase is inhibited by the end product isoleucine in *E. coli* (see Fig. 5.5).

Other metabolic situations may require both feedback activation and inhibition. Consider the pathways shown in Fig. 2.3. Substrate A goes into two pathways after being converted to C, which lead to the formation of roughly equal amounts of products G and N. To control the pathway so that G and N are kept in balance, high concentrations of G might inhibit the enzyme that converts C → D (feedback inhibition) and/or activate the enzyme that converts C → K (feedback activation). Conversely, N might inhibit the enzyme that converts C → K and/or activate the enzyme that converts C → D. G and N may inhibit the enzyme catalysing A → B to provide overall regulation. This line of control is found in the synthesis of purine nucleotides (see Fig. 5.7).

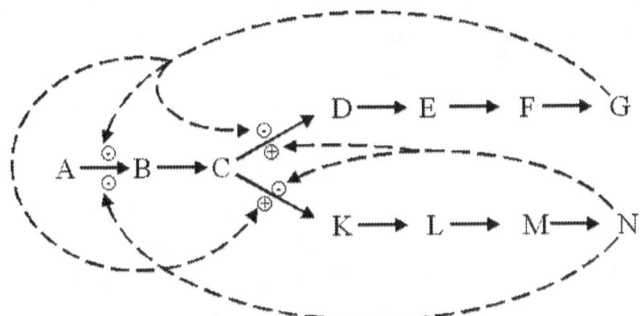

Fig. 2.3: Diagram showing substrate/product feedback control. (+) activation, (-) inhibition, (- - ▶) allosteric influence

Feed-forward activation

Feed-forward activation is a term used when a metabolite (e.g. B) produced earlier in a pathway activates an enzyme that catalyses a reaction (e.g. D to E) further down the pathway as shown below.

$$A \quad \cdot B \longrightarrow C \longrightarrow D \longrightarrow E \longrightarrow P$$

Allosteric Regulation

The activity of certain enzymes can be modulated by ligands. A ligand is any substance that is bound to a macromolecule. They can be activators or inhibitors of the enzyme. As indicated ealier those that cause a change in enzyme activity without being changed are called modulators or effectors. In addition to substrate binding sites (active sites) enzymes that respond to effectors have additional sites known as allosteric (regulatory) sites and are called

allosteric enzymes. In some allosterically regulated enzymes the catalytic sites and the regulatory sites are on separate polypeptide chains. Positive allosteric effectors (activators) increase the affinity of the enzyme for the substrate and other ligands whereas negative allosteric effectors (inhibitors) decrease the enzyme's affinity for the substrate or other ligands.

Most allosteric enzymes are oligomeric proteins with multiple ligand-binding sites, which change the behaviour of the enzymes through reversible changes in quaternary structure. Identical subunits are called **protomers.** Each protomer may consist of one or more polypeptide chains. Binding of a ligand to one protomer can affect the binding of similar ligands on other protomers in the oliogomer (e.g. substrate-substrate, inhibitor-inhibitor, activator-activator) in what is called **homotrophic** interactions. These interactions induce conformational changes that influence binding and catalytic activity at the active site and are almost always positive. On the other hand, the binding of one ligand may affect the binding of a different type ligand (e.g. inhibitor-activator; inhibitor-substrate; activator-substrate) in what is called **heterotrophic** interactions. Their effects are either negative or positive and can occur in monomeric allosteric enzymes. Both homotrophic and heterotrophic interactions involving oligomeric enzymes are mediated by cooperativity between subunits.

The binding of an allosteric effector causes allosteric transition in the enzyme (change in conformation of the enzyme) so that the affinity of the substrate or other ligands changes. They usually exhibit cooperativity where effects at one site positively or negatively affect the activity at the other sites.

Cooperativity

Cooperativity is defined as the influence that the binding of a ligand to one protomer has on the binding of the ligand to a second protomer of an oligomeric protein. It generally involves a change in conformation of an effector-activated protomer, which in turn transforms an adjacent protomer into a new conformation with an altered affinity for the effector ligand or for a second ligand.

The observation that most allosteric enzymes were oligomers and showed sigmoidal kinetics led to the concept of cooperativity to explain the interaction between ligand sites in oligomeric enzymes. However, kinetic mechanisms other than cooperativity can also produce sigmoidal plots. Hence sigmoidicity is not indicative of cooperativity in a V vs. [S] plot. One can have an allosteric effect in the absence of any cooperativity (i.e. induced conformational changes in one protomer are not transmitted to adjacent protomers).

Two models of cooperativity have been proposed: a) concerted and b) sequential-induced fit.

Concerted

The concerted model proposes that the enzyme exists in only two states: the T (Tense/Tight) and the R (Relaxed) states. These two states are in equilibrium (Fig. 2.4). Positive allosteric effectors favour the R state and shift the equilibrium towards the R state. Negative allosteric effectors favour the T state. A conformational change in one protomer causes a corresponding change in all protomers. The drawback of this model is that it cannot account for negative cooperativity

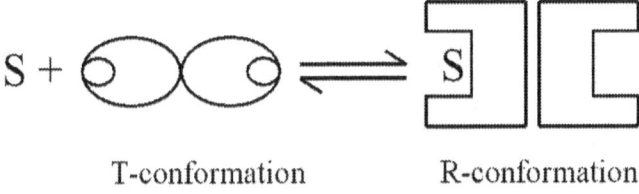

T-conformation   R-conformation

Fig. 2.4: The concerted model of cooperativity. Enzyme exists in two states the T (tense/tight) and the R (relaxed) conformations. Substrates and activators favour the R state and inhibitors favour the T state. S = substrate.

Sequential-induced fit

This model proposes that ligand binding to any one subunit (protomer) induces a conformational change in the protomer. This conformational change is transmitted particularly to adjoining protomers through protomer-protomer interactions. Thus, the effect of the first ligand binding is transmitted cooperatively and sequentially to the other protomers in the oligomer resulting in sequential increases or decreases in ligand affinity of the other protomers (Fig. 2.5). The cooperativity can be either positive or negative depending on the ligand.

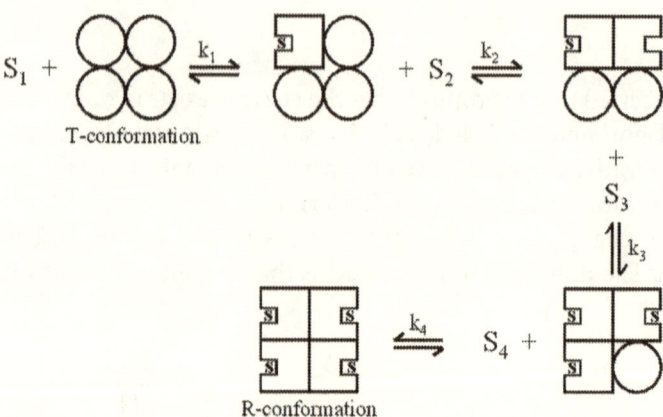

Fig. 2.5: The sequential-induced fit model of cooperativity. The binding of a ligand to any one subunit induces a conformational change in that subunit which is transmitted partially to adjoining subunits through subunit-subunit interactions. $S_1 \rightarrow S_4$ (substrates); $k_1 \rightarrow k_4$ (rate constants).

Sigmoidal kinetics of allosteric enzymes

As a result of the interaction between the substrate and activator or inhibitor sites characteristic sigmoidal or S-shaped curves are generally obtained in V vs. [S] plots of allosteric enzymes (Fig. 2.6). Negative allosteric effectors shift the curve to the right (towards high substrate concentrations) and increase the sigmoidicity of the curve (positive cooperativity). A high concentration of substrate will be required to achieve ½ V_{max} in the presence of a negative allosteric effector than is required in the absence of a negative effector. In the presence of a positive allosteric effector, ½ V_{max} can be reached at a lower substrate concentration than is required in the absence of a positive effector (negative cooperativity), i.e. the curve is shifted to the left and becomes closely hyperbolic.

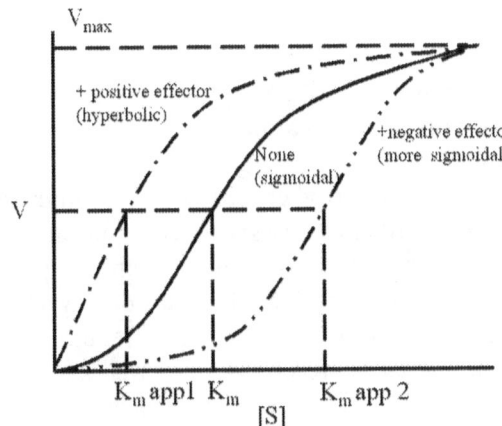

Fig. 2.6: Sigmoidal kinetics of a typical allosteric enzyme in the presence/absence of a positive/negative allosteric effector. Positive effector decreases the K_m ($K_m$app1)—negative cooperativity; negative effector increases the K_m ($K_m$app2)—positive cooperativity; V_{max} remains the same.

Allosteric enzymes allow fine control of the activity of individual enzymes, through small fluctuations in the concentration of substrate. Generally, *in vivo* concentrations of substrate correspond with the sharply rising segment of the sigmoidal plot. Consequently, large changes in enzyme activity are effected by small changes in substrate concentration. It is also possible to turn the enzymes off with small amounts of negative allosteric effector by the shifting of the apparent K_m to higher values, which are far above the *in vivo* concentrations of the substrate. At a given substrate concentration, the V is reduced in the presence of a negative effector.

Typical examples of such allosteric enzymes showing sigmoidal kinetics are (a) aspartate transcarbamoylase (ATCase) of *E. coli* the first allosteric enzyme to be thoroughly characterised (see chapter 5) and (b) PFK-1 (see chapter 3).

Covalent Modification

The activities of some regulatory enzymes are modulated by reversible covalent modification of the enzyme molecule. These include the phosphorylation, adenylylation, acetylation, uridylylation, ADP-ribosylation and methylation of the enzymes. The covalently attached groups are removed from the enzyme by separate enzymes.

The phosphorylation-dephosphorylation scheme of regulatory enzymes is the most common form of covalent modification found in eukaryotes. Some enzymes are phosphorylated on a single amino acid residue while others are phosphorylated at multiple sites. This type of covalent modification is central to a large number of regulatory pathways and will, therefore, be considered in greater detail.

The attachment of phosphoryl groups to specific amino acid residues of an enzyme is catalysed by protein kinases while the removal of phosphoryl groups is catalysed by protein phosphatases. The phosphoryl groups are attached to serine, threonine or tyrosine residues, thus introducing a bulky charged group into the enzyme. There are different cofactor requirements for the kinases involved in phosphorylation reactions. Two well-known examples are the cAMP-dependent and the Ca^{2+}-dependent protein kinases.

Cyclic AMP-dependent phosphorylation

The phosphorylation of an enzyme is a terminal result in a series of reactions initiated by the binding of a hormone to a specific extracellular target cell membrane receptor. For example, binding of epinephrine (ligand) to the β-adrenoceptor of plasma membrane causes a conformational change in the receptor leading to interaction between the receptor and a specific protein transducer, a guanine nucleotide binding (G) protein located on the cytosolic surface of the plasma membrane. The receptor-ligand complex activates the G protein, which in turn binds to and activates the effector enzyme adenylate cyclase (Fig. 2.7a).

Adenylate cyclase is a peripheral plasma membrane enzyme with its active site facing the cytosol. It catalyses the conversion of ATP to cAMP, a second messenger, which diffuses into the cytosol and causes the allosteric activation of a cAMP-dependent protein kinase (protein kinase A) by combining with its regulatory (R) subunit resulting in the formation of active catalytic (C) subunits (Fig. 2.7b). The activated cAMP dependent protein kinase can phosphorylate other inactive intracellular protein kinases and other enzymes and activate them. The phosphorylated form of a particular enzyme resulting from the protein kinase reaction may be either activated, e.g. glycogen phosphorylase (Fig. 3.10) or inactivated, e.g. glycogen synthase (Fig. 3.11). The mechanism by which a ligand like epinephrine does not enter the cell but through a cascade of signalling events from the surface of the plasma membrane produces an ultimate metabolic effect within the cell is called **signal transduction.**

A very important characteristic of most signalling pathways including the adenylate cyclase pathway, is **amplification.** It involves a cascade of reactions in which a single ligand-receptor complex can interact with and activate many G proteins resulting from rapid lateral diffusion of both receptors and G proteins within the membrane. A single activated molecule of adenylate cyclase may in turn produce many cAMP molecules. Four molecules of cAMP can produce two catalytically active protein kinase subunits. A single catalytically active protein kinase subunit can phosphorylate many target proteins. Thus, the signal generated by the binding of very low concentrations of hormone is multiplied many fold inside the cell.

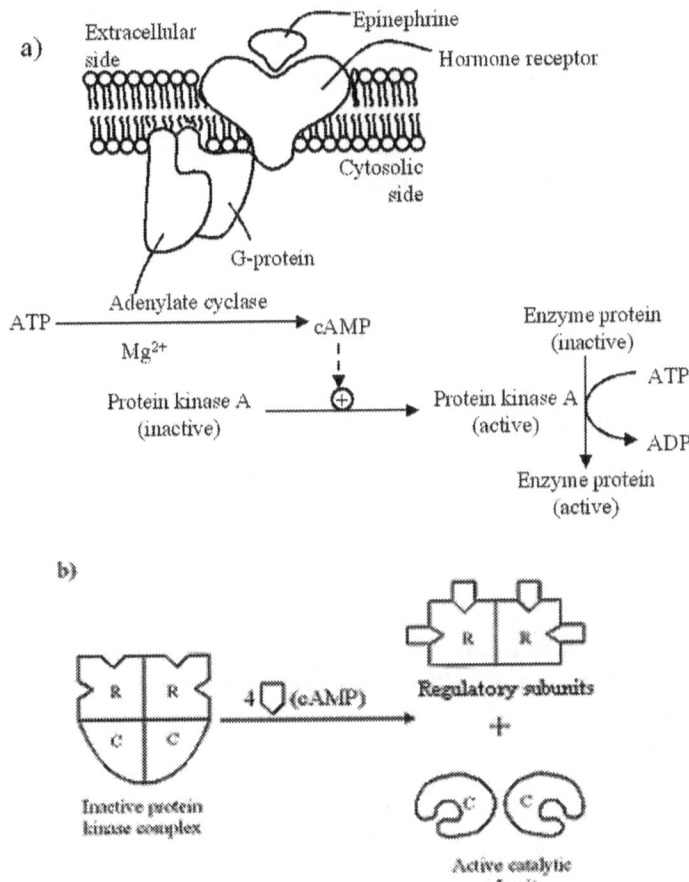

Fig. 2.7: Diagrams showing a) epinephrine activation of adenylate cyclase and b) cAMP-dependent activation of protein kinase. Mechanism of activation of protein kinase is briefly described in text. Adapted from Moran *et al* (1994). Biochemistry. Neil Patterson Publishers/Prentice Hall Inc., EnglewoodCliffs, New Jersey.

COARSE CONTROL

Long-term regulation of enzyme activity at the gene level is called coarse control. The rate of any reaction is dependent on the amount of enzyme present. Through the hormonal regulation of gene expression, the rates of synthesis of regulatory enzymes may be increased or decreased.

Insulin induces the synthesis of increased amounts of key regulatory enzymes such as glucokinase, phosphofructokinases 1 and 2 (PFK-1 and PFK-2), pyruvate kinase (PK) and glycogen synthase but represses the synthesis of gluconeogenic enzymes like PEP carboxykinase (PEPCK), pyruvate carboxylase (PC), glucose 6-phosphatase (G-6-Pase), fructose 1,6-bisphosphatase (F-1,6-BPase) and fructose 2,6-bisphosphatase (F-2,6-BPase).

Substances can sometimes repress the synthesis of enzyme protein, e.g. glucose represses the synthesis of PEPCK, the rate-limiting enzyme in the synthesis of glucose from pyruvate. The induction and repression of enzyme protein can be demonstrated by the lac and trp operons of *E. coli*, respectively. However, the detailed mechanisms of mammalian systems are unknown.

Control of Gene Expression for Catabolism

The lactose (lac) operon

The lac operon consists of three-linked structural genes that encode enzymes of lactose utilisation, and adjacent regulatory sites. The three structural genes: z, y and a, encode β-galactosidase (a hydrolytic enzyme), β-galactoside permease (a transport protein) and thiogalactoside transacetylase (an enzyme of unknown function), respectively (Fig. 2.8).

In the presence of an inducer, all three enzymes accumulate simultaneously but at different levels. Lactose itself causes the induction of the lactose operon and allolactose, the true intracellular inducer, is a minor product of β-galactosidase action.

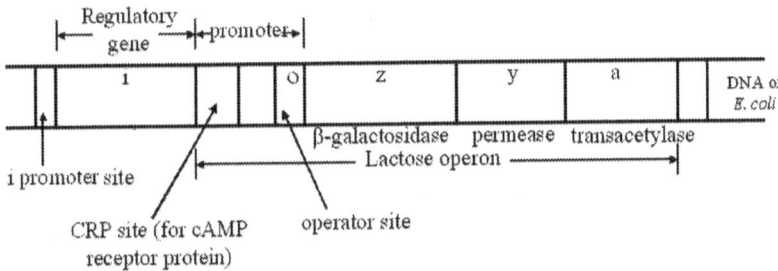

Fig. 2.8: The lac operon of *E. coli*. It comprises of the *i* gene, which encodes the lac repressor and the *o* site to which the repressor binds. The lac *z*, *y* and *a* are the structural genes that encode β-galactosidase, β-galactoside permease and thiogalactoside transacetylase, respectively.

Regulation of lac operon

Jacob-Monod model

Transcription of the three structural genes is initiated near an adjacent site, the operator, and yields a single polycistronic messenger RNA, an RNA copy of the three genes (cistron—a region of a genome that encodes one polypeptide chain). The regulatory gene product is a repressor, which can exist in active and inactive forms. In the active form, it reacts with the operator site and blocks the transcription of the structural genes. The repressor on the other hand can interact with the inducer (a low molecular weight molecule), which converts it into the inactive form (repressor-inducer complex) that cannot bind to the operator site and, therefore, allows the transcription of the structural genes in a process called **induction** or **derepression** (Fig. 2.9a, b).

In the lac operon, the binding of isopropyl thiogalactoside (IPTG), allolactose, or some other inducer to the repressor inactivates the repressor by decreasing its affinity for the operator. This repressor inactivation stimulates transcription of z, y and a, because dissociation of the repressor-inducer complex from the operator removes a steric block to the binding of RNA polymerase at the initiation site. Thus, the introduction of lactose or a similar inducer activates synthesis of the gene products involved in its catabolism by removing the barrier to their transcription. This mode of regulation is essentially negative, because the active regulatory element (the repressor) is an inhibitor of transcription.

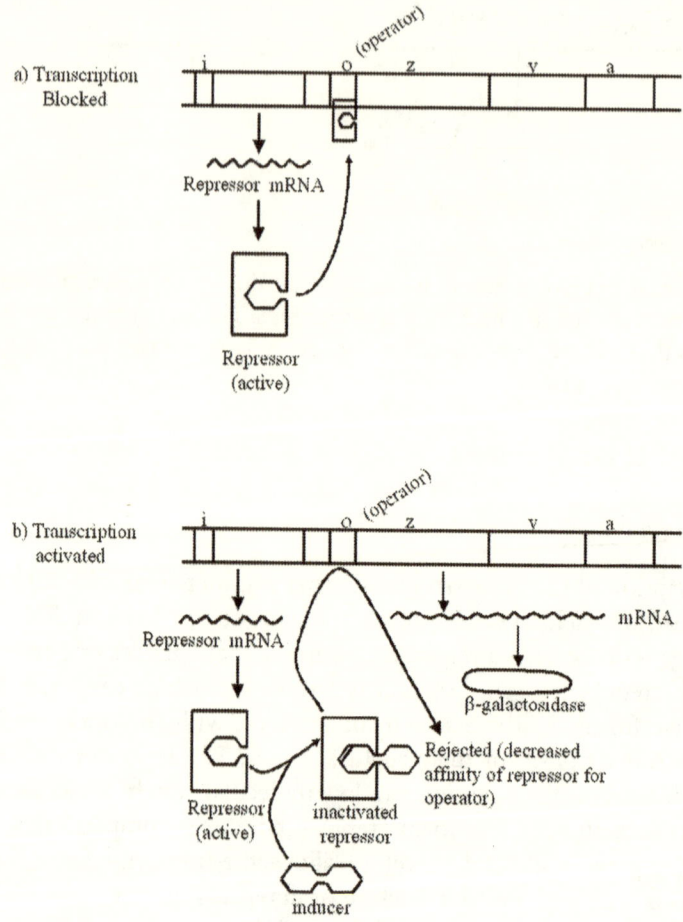

Fig. 2.9: Regulation of lac operon. a) The repressor (active) binds to the operator and prevents the transcription of the structural genes; b) the inducer binds to the repressor to form the inducer-repressor complex (inactive), which prevents the repressor from binding to the operator. This causes the transcription of the structural genes from the promoter site.

Catabolite repression by glucose

The repressor-operator system keeps the operon turned off in the absence of the utilizable β-galactosides. An overlapping regulatory system turns the operon on only when alternative energy sources are unavailable. This is a positive control system. When *E. coli* is grown in a medium containing glucose and lactose, the cells metabolise glucose exclusively until the supply is exhausted. Then growth slows, and the lac operon is activated in preparation for continued growth using lactose. This phenomenon, now known to involve a transcriptional activation mechanism, was originally called glucose repression or catabolite repression. Transcriptional activities occur when glucose levels are low, and control is exerted through intracellular levels of cAMP (Fig. 2.10).

The cAMP levels in *E. coli* are low when intracellular glucose levels are high. The mechanism of regulation is not fully known but it appears adenylate cyclase senses the intracellular levels of an unidentified intermediate in glucose catabolism (catabolite activation). When glucose levels drop, cAMP levels trigger the activation of the lac operon by its interaction with a protein called cAMP receptor protein (CRP) formerly called catabolite activator protein (CAP). When CAP binds cAMP, it undergoes a conformational change that increases its affinity for certain DNA sites including a site (CAP site) in the lac operon adjacent to the RNA polymerase binding site. This in turn facilitates transcription of the lac operon by stimulating the binding of RNA polymerase to form a closed promoter complex.

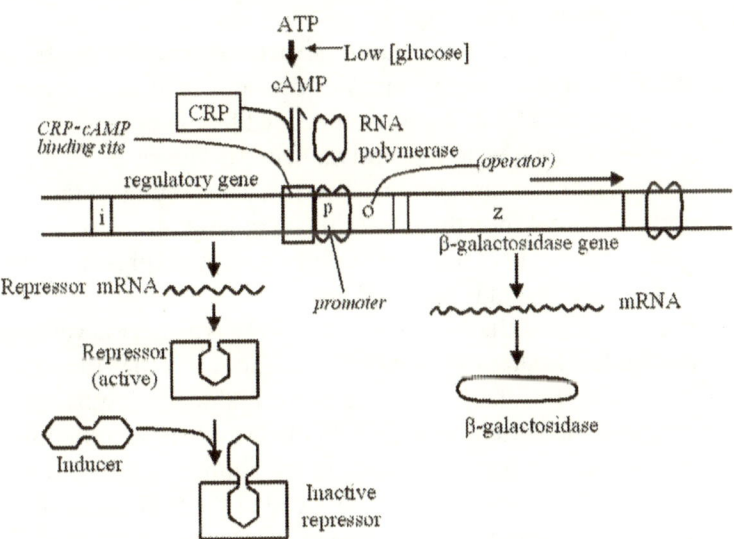

Fig. 2.10: Regulation of the lac operon by cAMP. When glucose levels are low (or at high cAMP levels) binding of CAP (CRP)-cAMP complex to the CAP site on DNA adjacent to the RNA polymerase binding site (the promoter) causes stimulation of binding of RNA polymerase leading to increased transcription of the lac operon. At low cAMP levels the complex is not formed and, therefore, transcription is curtailed.

Control of Gene Expression for Biosynthesis

The tryptophan (trp) operon

The lac operon is involved in the catabolism of a substrate and, therefore, the gene products are not needed unless the substrate is also present to be consumed. A different situation is encountered with genes whose products catalyse the biosynthesis of amino acids such as trp. Biosynthesis consumes energy and, therefore, it is in the interest of the cell to use preformed amino acids, if available. Therefore the regulatory goal is to repress gene activity by turning off the synthesis of enzymes in the pathway when the end product is available.

Regulation of the trp operon

Regulation of the trp operon of *E. coli* demonstrates two ways of accomplishing this shutdown.

a) A repressor design in which binding of a small molecule or ligand activates the repressor.

b) Early termination of transcription.

The trp operon consists of five adjacent structural genes whose transcription is controlled from a common promoter-operator regulatory region (Fig. 2.11). The trp repressor, a 58 kDa protein encoded by the non-adjacent trp R gene, binds a low molecular weight ligand namely trp. However, the repressor-ligand complex is the active form of the repressor, which binds to the operator and blocks transcription. When the intracellular trp levels decrease, the repressor-ligand complex dissociates and the free protein (aporepressor) leaves the operator so that transcription is activated.

Charles Yanofsky found that the activation of the trp enzyme varies over a 600-fold range under different physiological conditions, more than could be accounted for by a repressor-operator mechanism alone. Analysis revealed a second mechanism called **attenuation**, which involves early termination of the trp operon transcription under conditions of trp abundance. A site called the attenuator ("a") is 131 nucleotides from the 5' end of the trp L sequence. When trp levels are high, transcription terminates at "a", to give an attenuated 131+ nucleotide transcript rather than the complete 7,000 nucleotide trp mRNA. The structural genes are not transcribed, and, therefore, trp is not synthesised.

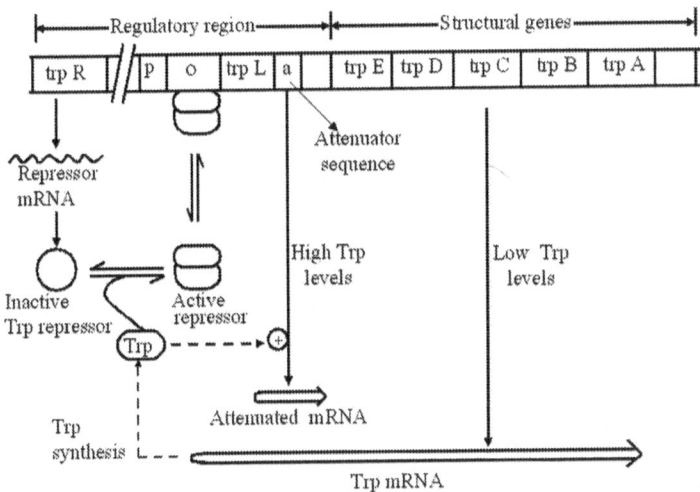

Fig. 2.11: The trp operon of *E. coli*. It is made up of five adjacent structural genes. At high intracellular trp levels the formation of trp-repressor complex causes the activation of the repressor. The binding of the activated repressor to the operator leads either to a blockage or the attenuation of transcription. At low levels of trp, the ligand-repressor complex dissociates and the free repressor protein (aporepressor) leaves the operator leading to the activation of transcription.

Further reading

Cleland, W.W. (1977). Determining the Chemical Mechanisms of Enzyme-catalyzed Reactions by Kinetic Studies. *Adv. Enzymol.* 45:273-387

Devlin, T.M. (1992). Textbook of Biochemistry with Clinical Correlations. Wiley-Liss Inc., New York

Dische, Z. (1976). The Discovery of Feedback Inhibition. *Trends Biochem. Sci.* 1:269-270

Koshland, D.E., Jr. and Neet, K.E. (1968). The Catalytic and Regulatory Properties of Enzymes. *Annu. Rev. Biochem.* 37:359-410

Matthews, C.K. and van Holde, K.E. (1996). Biochemistry. The Benjamin/Cummings Publishing Co. Inc., Merlo Park, California

Moran, L.A., Scrimgeour, K.G., Horton, R.H., Ochs, R.S., and Rawn, J.D. (1994). Biochemistry. Neil Patterson Publishers/Prentice Hall Inc., Englewood Cliffs, New Jersey

PART TWO
CONTROL OF CELLULAR METABOLISM

CHAPTER THREE

CARBOHYDRATE METABOLISM

The regulation of carbohydrate metabolism includes both anabolic and catabolic pathways.

CATABOLISM

Carbohydrate catabolism provides energy in the form of ATP and other metabolic precursors for biosynthetic processes. It is important that the concentrations of these substances in the body are kept relatively constant irrespective of which fuel is being used or how energy is being expended. There are four different pathways of carbohydrate catabolism: i) glycolysis, ii) citric acid cycle (CAC), iii) glycogenolysis and iv) pentose phosphate pathway (PPP) which is also classified as an anabolic pathway. The pathway that predominates depends on three main factors namely:

Nutritional status: In the fasted state glycogenolysis predominates as the body tries to attain normal blood glucose levels.

Oxygen status :   In very active muscle when oxygen demand exceeds supply, glycolysis predominates as the main source of energy since activity of the electron transport chain (ETC) is slowed down.

Tissue type :   The PPP is prominent in adipose tissue since NADPH, a major product of the pathway, is required in FA synthesis.

ANABOLISM

The biosynthesis of carbohydrates includes the synthesis of glucose from gluconeogenic precursors like pyruvate and lactate and storage carbohydrates like glycogen from glucose. There are two main pathways of carbohydrate biosynthesis: glycogenesis and gluconeogenesis. The pathway that predominates depends on the nutritional status and tissue type. Thus, in a well-fed state excess glucose is stored as glycogen in the liver and muscle, and fatty acids (FAs) as TAGs in the adipocytes.

REGULATION OF CATABOLIC PATHWAYS

GLYCOLYSIS

There are basically three enzymes responsible for the regulation of the glycolytic pathway. These are (a) glucokinase or hexokinase depending on the tissue, (b) phosphofructokinase-1 (PFK-1) and (c) pyruvate kinase (PK). These enzymes catalyse the three regulatory steps shown in Fig. 3.1. Glycolysis has more than one regulatory point because it not only generates ATP and provides pyruvate for oxidation via the CAC but also provides intermediates for other biosynthetic pathways. For example, dihydroxyacetone phosphate (DHAP) and 3-phosphoglycerate are used in the synthesis of lipids and amino acids, respectively.

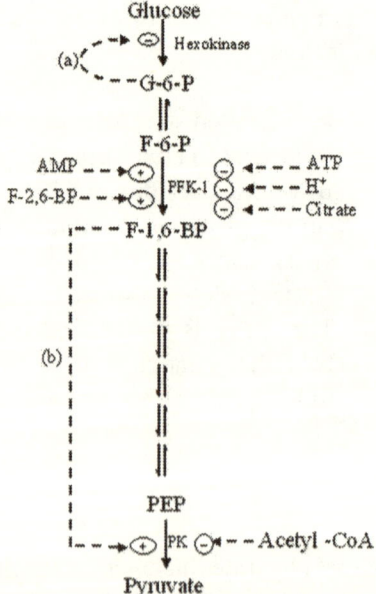

Fig 3.1: **Allosteric regulation of the glycolytic pathway.** (-) inhibition; (+) activation; (- - ▶) allosteric influence; (a) feedback inhibition; (b) feed-forward activation.

Hexokinase/Glucokinase

Hexokinase, which is found in most tissues except liver and pancreas, has a low K_m (0.10 mM) for glucose compared to the blood glucose concentration of 5 mM. It is strongly inhibited by the product of its reaction, G-6-P. Glucokinase, an isozyme of hexokinase found in the pancreas and liver, has a higher K_m of 10 mM (twice the normal blood glucose concentration of 5 mM) and is not subject to inhibition by G-6-P. This high K_m of glucokinase helps the liver to regulate blood glucose levels, since at blood glucose concentration of 5 mM glucokinase is not saturated whereas hexokinase is (Fig. 3.2). An increase in glucose concentration above 5 mM leads to a proportionate increase in the rate of glucose phosphorylation by glucokinase while a reduction leads to a corresponding decrease.

The low K_m of hexokinase allows glucose to be phosphorylated even at concentrations below normal physiological blood/tissue glucose levels. This is good particularly for the brain, which depends largely on glucose even when blood and tissue glucose levels are dangerously low.

Fine control

Allosteric

Hexokinase undergoes allosteric control by the product of its reaction, G-6-P, which acts as a negative effector of the enzyme, in a process called product inhibition or feedback inhibition.

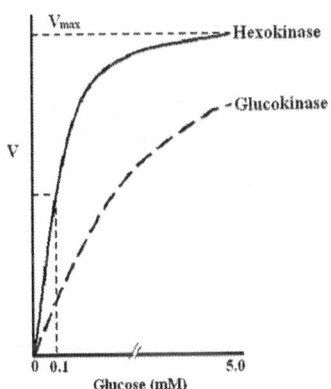

Fig. 3.2: Kinetic features of hexokinase and glucokinase

Coarse control

Glucokinase is subject to induction/repression of synthesis under hormonal control. In the presence of insulin, the amount of glucokinase is increased through the promotion of transcription of the glucokinase gene and vice versa when glucagon is predominant. Thus, the amount of glucokinase present is a reflection of the amount of glucose reaching the liver. Under normal physiological conditions, the higher the carbohydrate content of the meal the higher the levels of insulin and glucokinase.

Phosphofructokinase-1 (PFK-1)

The conversion of F-6-P to F-1,6-BP, catalysed by PFK-1, is the second and most important site of regulation in the glycolytic pathway. This enzyme catalyses the committed step of the glycolytic pathway and is subject to the greatest degree of fine control (Figs 3.3 and 3.4).

Allosteric control

The kinetics of allosteric regulation of PFK-1 is shown in Fig. 3.3.

Negative effectors

ATP   :   It is both a substrate and in most cases an allosteric inhibitor of PFK-1. At high concentrations it causes a decrease in the affinity of PFK-1 for F-6-P thus increasing the K_m of the enzyme ($K_m$app1) for F-6-P (Fig. 3.3).

Citrate :   Oxidation of FAs and ketone bodies causes the elevation of cytosolic levels of citrate, which is an allosteric inhibitor of PFK-1. It causes a decrease in the glucose utilisation by tissues when FAs and ketone bodies are readily available. This represents a feedback inhibition that regulates the supply of pyruvate to the CAC.

H^+ : H^+ is a glycolytic end product (from lactic acid), which inhibits PFK-1 and shuts off glycolysis to prevent lactic acidosis.

Positive effectors

AMP   :   When ATP levels increase, AMP levels decrease and vice versa. Considering the reaction:

$$2\ ADP \rightleftharpoons ATP + AMP$$

$$K_{eq} = \frac{[ATP]\,[AMP]}{[ADP]^2} \qquad [AMP] = \frac{K_{eq}\,[ADP]^2}{[ATP]}$$
,

Since intracellular $[ATP] \gg [ADP] \gg [AMP]$ under normal physiological conditions, and total cellular concentration of these three substances is constant, a small decrease in [ATP] causes a greater percentage increase in [ADP], and an even greater percentage increase in [AMP]. This makes AMP an excellent signal of the energy status of the cell and allows it to function as an important positive allosteric effector of PFK-1 activity, thus decreasing the K_m of the enzyme ($K_m$app2) for F-6-P (Fig. 3.3).

Fructose 2,6-Bisphosphate : F-2,6-BP is formed from F-6-P by the action of phosphofructokinase-2 (PFK-2), which is stimulated by P_i and inhibited by citrate. A different active site of the same enzyme protein (bifunctional enzyme) catalyses the dephosphorylation of F-2,6-BP to F-6-P, the enzyme being F-2,6-BPase. In the presence of F-2,6-BP, PFK-1 is activated (i.e. its K_m for F-6-P is decreased; $K_m$app2) as shown in Fig. 3.3, and F-1, 6-BPase is inhibited. In its absence, PFK-1 is inhibited and F-1,6-BPase is activated preventing net flux through the glycolytic pathway.

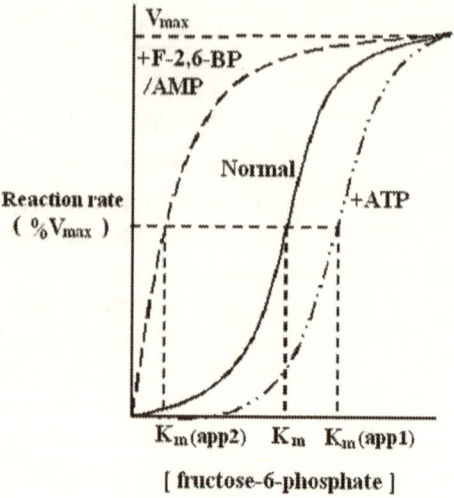

[fructose-6-phosphate]

Fig. 3.3 Kinetics of allosteric regulation of liver PFK-1 by F-2,6-BP/AMP (positive effectors) and ATP (negative effector). F-2,6-BP/AMP lowers K_m ($K_m$app2) while ATP increases the K_m ($K_m$app1) of PFK-1 for F-6-P.

Covalent modification

The bifunctional enzyme PFK-2/F-2,6-BPase is able to carry out both the synthesis and degradation of F-2,6-BP. It is subject to cAMP-dependent phosphorylation and dephosphorylation, which is under hormonal control because hormones influence the intracellular cAMP concentration.

In the liver, dephosphorylation leads to activation of PFK-2, while phosphorylation causes activation of F-2,6-BPase. The nutritional status of the individual determines the predominant circulating hormone, which controls the intracellular concentration of cAMP. Thus, the activity of the liver bifunctional enzyme is closely tied to the physiological state of the individual.

It has been reported that the bifunctional enzyme in the muscle is an isoform of that in the liver and cAMP has an opposite effect. In muscle, both insulin and epinephrine activate PFK-2 through its phosphorylation. Epinephrine—induced activation of PFK-2 by cAMP-dependent protein kinase causes a decrease in the K_m of the enzyme for F-6-P. In contrast, stimulatory effect of insulin on PFK-2, which appears to be mediated by protein kinase B, does not alter the K_m but increases V_{max}.

Fed state

In the fed state insulin is predominant and so cAMP levels are relatively low. Thus, the ratio of blood insulin/glucagon determines the intracellular levels of F-2,6-BP in the liver which influence the rate of liver glycolysis. Dephosphorylation of the bifunctional enzyme as a result of low cAMP levels inactivates F-1,6-BPase and makes PFK-2 more active. This increases the production of F-2,6-BP with the consequent stimulation of PFK-1 to increase flux through the glycolytic pathway (Fig. 3.4).

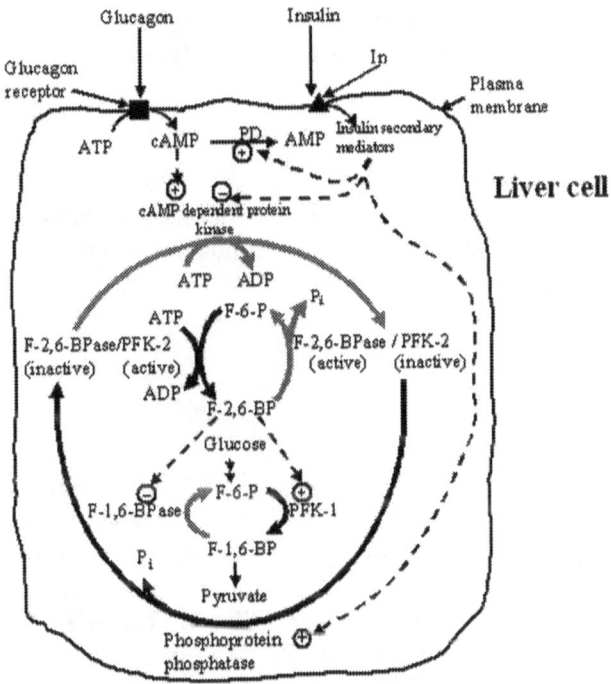

Fig. 3.4: Hormonal control of the bifunctional enzyme PFK-2/F-2,6-BPase and consequent effects on PFK-1 and F-1,6-BPase in liver. ▬▬▬ Glucagon actions; ▬▬▬ Insulin actions. PD = c AMP phosphodiesterase; (-) inhibition; (+) activation; (- - ▶) allosteric influence.

Starved state

In a starved state, glucagon is predominant since there is need to mobilize liver glycogen and enhance gluconeogenesis to maintain glucose homeostasis. Hence, flux through the glycolytic pathway is significantly reduced. Glucagon triggers adenylate cyclase of liver plasma membrane to produce cAMP from ATP. The cAMP in turn activates a protein kinase to phosphorylate the bifunctional enzyme PFK-2/F-2,6-BPase making PFK-2 inactive and F-2,6-BPase active. This causes a fall in F-2,6-BP levels with the consequent reduction in PFK-1 activity. Thus, conversion of F-6-P to F-1,6-BP is significantly reduced and flux through the glycolytic pathway is curtailed (Fig. 3.4). This makes glucagon an extracellular signal that stops the liver from using glucose whereas [F-2,6-BP] is an intracellular signal that promotes glucose utilisation.

Pyruvate Kinase

The conversion of PEP to pyruvate, catalysed by the enzyme pyruvate kinase (PK), is the third regulatory site in the glycolytic pathway. However, like the hexokinase/glucokinase step, it is a secondary site of regulation. Regulation involves both fine and coarse controls.

Fine control

Allosteric

ATP   :   Like PFK-1, PK is drastically inhibited by high concentrations of ATP (K_m of PK for PEP is increased; $K_m$app2). High ATP levels reduce the apparent affinity of PK for PEP (Fig. 3.5).

F-1, 6-BP :   Produces a feed-forward activation of PK (Fig.3.1). Since F-1,6-BP is an activator of PK, activation of PFK-1 causes subsequent activation of PK (K_m of PK for PEP is decreased; $K_m$app1). This ensures that intermediates passing the first committed step of glycolysis at PFK-1 are able to pass through the pathway without getting accumulated. Figure 3.5 also indicates that at a given concentration of PEP, PK activity is greater in the presence of the allosteric activator F-1,6-BP up to the saturating concentrations of PEP.

Acetyl-CoA: Acts as an allosteric inhibitor of PK (Fig. 3.1). This occurs when there is excessive production of acetyl-CoA through the breakdown of fat, reducing flux through the glycolytic pathway.

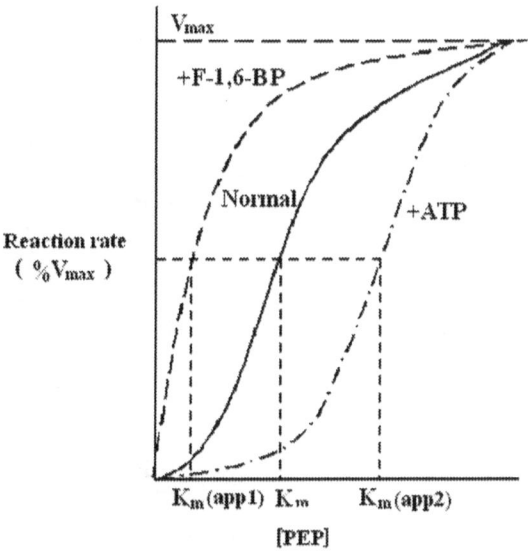

Fig. 3.5: Kinetics of the allosteric regulation of PK by F-1,6-BP and ATP. The K_m of PK for PEP is decreased (K_m app1) in the presence of F-1,6-BP but increased ($K_m$app2)in the presence of ATP.

Covalent modification

Isozymes of PK found in mammalian liver and intestinal cells undergo covalent modification. The phosphorylation and consequent inactivation of PK is catalysed by protein kinase A. Dephosphorylation of PK by a phospho-protein phosphatase renders it active (Fig. 3.6). The change in kinetic behaviour of PK on phosphorylation is demonstrated by the fact that addition of glucagon to liver cells shifts the sigmoidal curve to the right (K_m app > K_m) suggesting that phosphorylation reduces the activity of PK (Fig. 3.7).

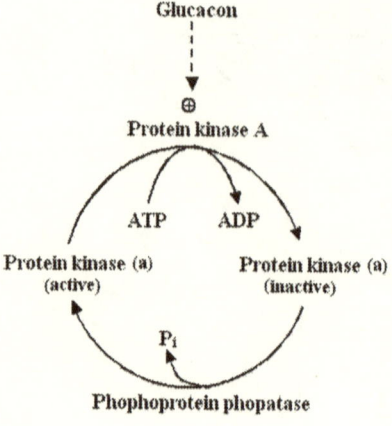

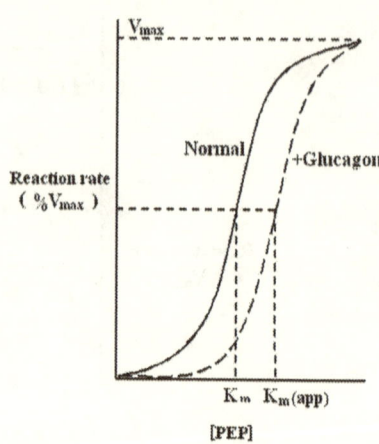

Fig. 3.6: Covalent modification of PK activity by glucagon.

Fig. 3.7: Effects of glucagon on the kinetics of PK. Glucagon-induced phosphorylation increases the K_m of PK for PEP ($K_m$app).

Coarse control

A high glucagon/insulin ratio causes a repression of the synthesis of PK leading to reduced glycolysis and the stimulation of gluconeogenesis.

PYRUVATE DEHYDROGENASE COMPLEX

The regulation of the pyruvate dehydrogenase (PDH) complex, which catalyses the conversion of pyruvate to acetyl-CoA for its entry into citric acid cycle (CAC) is shown in Fig. 3.8.

Fine control

Allosteric

- The transacetylase component (E_2) is inhibited by acetyl-CoA and activated by CoASH.

- The dihydrolipoamide dehydrogenase component (E_3) is inhibited by NADH and activated by NAD^+.

- ATP is an allosteric inhibitor of the complex while AMP is an activator.

Thus, the ratio of [NAD+]/[NADH] and that of [acetyl-CoA]/[CoASH] determine the activity of the enzyme complex.

Covalent modification

The mammalian PDH complex undergoes covalent modification of the pyruvate dehydrogenase component of the complex (E_1). This involves the phosphorylation and dephosphorylation of a specific serine residue on one of the two subunits of E_1.

Inactivation of the PDH complex is accomplished by a Mg^{2+}-ATP-dependent protein kinase (PDH kinase), which phosphorylates the enzyme complex. The reactivation of the complex is accomplished by a phosphoprotein phosphatase (PDH phosphatase), which dephosphorylates the complex in a Mg^{2+}/Ca^{2+}-dependent enzymatic reaction. Thus, the differential regulation of the PDH kinase and the PDH phosphatase is the key to the regulation of the PDH complex.

The essential features of this complex regulatory system involves a combination of allosteric control and covalent modification as follows:

- NADH and acetyl-CoA do not only inhibit the dephosphorylated form (active) of the PDH complex but also activate PDH kinase leading to the phosphorylation and inactivation of the PDH complex.

- Free CoASH and NAD^+ inhibit the PDH kinase, thus activating the PDH complex. Hence any increase in mitochondrial $NADH/NAD^+$ or acetyl-CoA such as during rapid β-oxidation of FAs will inactivate PDH through stimulation of the PDH kinase.

- Pyruvate is a potent inhibitor of PDH kinase and, therefore, in the presence of elevated tissue pyruvate levels, the kinase is inhibited and the PDH complex maximally activated.

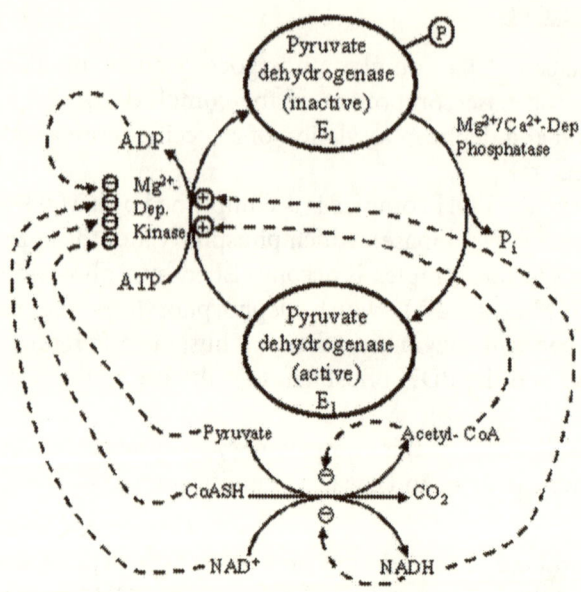

Fig. 3.8: Diagram showing allosteric and covalent modification regulations of pyruvate dehydrogenase complex. (-) inhibition; (+) activation; (- - ▶) allosteric influence. Mg^{2+}/Ca^{2+} Dep. = magnesium/calcium dependent

CITRIC ACID CYCLE (CAC)

The passage of carbon atoms of pyruvate into and through the CAC is controlled at two levels.

a) The conversion of pyruvate to acetyl-CoA for entry into the cycle as shown in Fig. 3.8.

b) Regulation by key enzymes in the CAC under the allosteric influence of cofactors and intermediates.

Three factors govern the rate of flux through the cycle.

- Substrate availability (oxaloacetate, acetyl-CoA and NAD^+)
- Inhibition by accumulating products (succinyl-CoA, citrate and ATP)
- Allosteric feedback inhibition of enzymes catalysing the early steps of the cycle.

The regulation is at the three exergonic steps in the cycle catalysed by: 1) citrate synthase, 2) isocitrate dehydrogenase and 3) α-ketoglutarate dehydrogenase (Fig. 3.9).

Substrate availability: The availability of substrates for citrate synthase (acetyl-CoA and oxaloacetate) depends on the metabolic state of the cell and may limit the rate of citrate formation.

Product accumulation: Inhibits all the three limiting steps of the cycle.
Succinyl-CoA—inhibits α-ketoglutarate dehydrogenase and citrate synthase.
Citrate—inhibits citrate synthase.
NADH—a product of isocitrate and α-ketoglutarate dehydrogenases activity accumulates under certain conditions leading to the inhibition of both enzymes at high NADH/NAD$^+$ ratio.
ATP—inhibits citrate synthase and isocitrate dehydrogenase.

Respiratory control: The activity of the respiratory chain is tightly coupled to the generation of ATP in oxidative phosphorylation, which depends on the availability of ADP, P$_i$ and O$_2$. Thus, anything that reduces the supply of O$_2$, ADP and reducing equivalents shuts or slows down the respiratory chain, and consequently, the CAC.

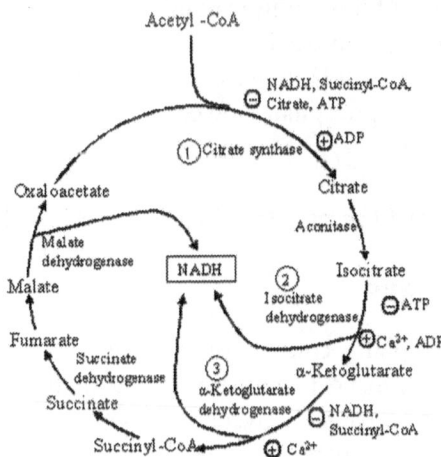

Fig. 3.9: Regulation of key enzymes of the citric acid cycle (CAC). (-) inactivation; (+) activation

PENTOSE PHOSPHATE PATHWAY

The pentose phosphate pathway (PPP) or hexose monophosphate shunt (HMS) is active in tissues that synthesize FAs and steroids, e.g. mammary gland, liver, adrenal gland and adipose tissue, since large amounts of NADPH are consumed in these biosynthetic processes. In other cells like muscle and brain, the PPP accounts for little of the overall consumption of glucose. It is present in erythrocytes for NADPH production, which is used to generate reduced glutathione essential for the maintenance of RBC structure. Enzymes catalysing the reactions of this pathway are found in the cytosol, site of the many biosynthetic reactions that require NADPH.

This pathway can be divided into oxidative and non-oxidative phases with two primary functions: i) to provide NADPH for reductive biosynthetic processes and ii) to provide ribose 5-phosphate (ribose 5-P) for nucleotide biosynthesis. The actual fate of G-6-P after entry into the pathway depends on the metabolic needs of the cell in which the pathway occurs.

Oxidative Phase

The first step in the oxidative phase; the conversion of G-6-P to 6-phospho-gluconolactone catalyzed by glucose 6-phosphate dehydrogenase (G-6-PD) is the major regulatory site for the entire PPP. NADPH allosterically inhibits G-6-PD, and thus, the production of NADPH as G-6-P is converted to ribulose 5-phosphate (ribulose 5-P) is self-limiting.

If the cells carrying out this pathway require large amounts of both NADPH and nucleotides, all the ribulose 5-P is converted to ribose 5-P to complete the pathway. Usually more NADPH than ribose 5-phosphate is needed and, there-fore, most of the pentose phosphates are converted to glycolytic intermediates in the non-oxidative phase.

$$G\text{-}6\text{-}P + 2\,NADP^+ + H_2O \longrightarrow Ribulose\ 5\text{-}P + 2\,NADPH + CO_2 + 2\,H^+$$

Non-oxidative Phase

The non-oxidative phase consists of entirely near-equilibrium reactions and provides a means of disposing of pentose phosphates formed in the oxida-tive phase by providing a route to glycolysis or gluconeogenesis. Ribulose 5-P is converted to F-6-P and glyceraldehyde 3-P, intermediates of the glycolytic pathway. If more ribose 5-P than NADPH is required the reversal of this process occurs.

3 Ribulose 5-P $\longrightarrow$ 2 F-6-P + glyceraldehyde 3-P

GLYCOGENOLYSIS

Glycogenolysis is the catabolism of glycogen, which is catalysed by the enzyme glycogen phosphorylase. Glycogen phosphorylase is a regulatory enzyme and is, therefore, subject to fine control by allosteric effectors and covalent modification (Fig. 3.10). It catalyses the first step in the degradation of glycogen. This occurs in a fasted state or under stressful conditions when glucagon or epinephrine levels are high with correspondingly low levels of insulin.

Regulation of Glycogen Phosphorylase

Allosteric control

Glycogen phosphorylase is subject to allosteric activation by AMP and inhibition by ATP only in the muscle, but allosteric inhibition by glucose only in the liver. This forms a minor part of the control process.

Covalent modification

Glycogen phosphorylase undergoes very elaborate control by covalent modification. The enzyme exists in two forms; the active 'a' form and the relatively inactive 'b' form. These forms of the enzyme undergo interconversion by the actions of phosphorylase kinase and phosphoprotein phosphatase-1 (PP-1) as described below.

Phosphorylation transforms the relatively inactive glycogen phosphorylase 'b' to the active glycogen phosphorylase 'a' and the reverse is true during dephosphorylation. However, there is the allosteric mechanism for partially activating glycogen phosphorylase 'b' and inactivating glycogen phosphorylase 'a'; AMP activates the relatively inactive glycogen phosphorylase 'b' but has little effect on the already active glycogen phosphorylase 'a', while glucose or ATP inhibit active glycogen phosphorylase 'a' but are without effect on relatively inactive glycogen phosphorylase 'b'. Thus, covalent modification can be supplemented by allosteric control and vice versa.

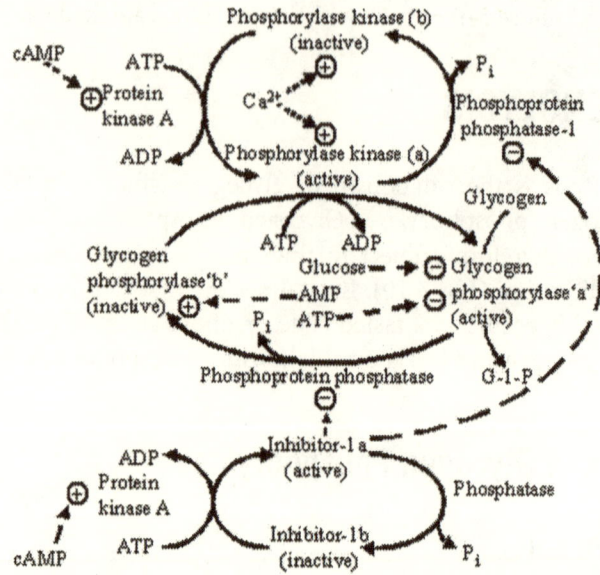

Fig. 3.10: Regulation of glycogen phosphorylase by covalent modification and allosteric influence. (-) inactivation; (+) activation; (- - ▶) allosteric influence.

Phosphorylase kinase

Phosphorylase kinase is responsible for the phosphorylation and activation of glycogen phosphorylase. However it is also subject to regulation by a cyclic phosphorylation-dephosphorylation mechanism. The cAMP-dependent protein kinase (protein kinase A or PKA) is responsible for the phosphorylation and activation of phosphorylase kinase and PP-1 is in turn responsible for the dephosphorylation and inactivation of phosphorylase kinase.

The enzyme consists of four subunits with four molecules/subunit in the complex $\alpha_4\beta_4\gamma_4\delta_4$. Catalytic activity is associated with the γ-subunit while the α,β and δ subunits exert regulatory control. Phosphorylation of one serine residue in each of the α and β-subunits converts the inactive form of the enzyme phosphorylase kinase 'b' to the active form phosphorylase kinase 'a', which then activates glycogen phosphorylase from the inactive 'b' form to the active 'a' form.

The δ subunit is a Ca^{2+} binding regulatory protein called calmodulin, which is not unique to phosphorylase kinase. It functions as a Ca^{2+} receptor in cells responding to changes in intracellular Ca^{2+} concentration and modulates the activities of some enzymes. The binding of Ca^{2+} to the calmodulin subunit of phosphorylase kinase changes the conformation of the complex making the enzyme more active with respect to the phosphorylation of glycogen phosphorylase.

Phosphoprotein phosphatase-1

The ultimate control of glycogen phosphorylase activation involves the simultaneous turning off of PP-1 and turning on of phosphorylase kinase and vice versa for the inactivation of the enzyme. Since PP-1 also acts on phosphorylase kinase, turning off PP-1 would also cause a greater activation of phosphorylase kinase.

Inhibitor-1 (I-1), a protein contained in cells, inhibits PP-1. This protein is subject to covalent modification by PKA and PP-1. Only the 'a' form (phosphorylated) of phosphatase inhibitor-1 (I-1a) inhibits PP-1 and phosphorylation occurs on a threonine instead of a serine residue. The I-1a is converted back to the dephosphorylated 'b' form (I-1b) by a phosphatase.

This bicyclic control system for the activation and inactivation of glycogen phosphorylase provides a tremendous amplification mechanism (Fig. 3.10).

REGULATION OF ANABOLIC PATHWAYS

GLYCOGENESIS

The regulatory enzyme in glycogen synthesis is glycogen synthase. It is a glycosyltransferase that transfers an activated sugar unit (UDP-glucose) to a non-reducing sugar hydroxyl group generating an α $(1\rightarrow 4)$ glycosidic linkage between carbon 1 (C_1) of the incoming glycosyl moiety and the carbon 4 (C_4) of the glucose residue at the terminus of the glycogen chain. The enzyme successively adds glucose units to the 4-hydroxyl group at the non-reducing end.

Glycogen Synthase

Glycogen synthase is a tetrameric protein made up of identical α-subunits. It exists in the phosphorylated and dephosphorylated states, with the phosphate reversibly bound to a serine residue on each subunit. The dephosphorylated form of glycogen synthase; glycogen synthase 'a' is the active form, which is independent of G-6-P for activity. The phosphorylated form is the less active glycogen synthase 'b', which is dependent on G-6-P for activity. Glycogen synthase undergoes fine control mechanisms (Fig. 3.11).

Fine Control

Allosteric

When G-6-P concentrations are high, it acts as a positive allosteric effector of glycogen synthase 'b' leading to activation of glycogen synthase so that it signals the conversion of some of the G-6-P into glycogen. Glycogen is able to exert negative feedback control over its own synthesis. It may either activate protein kinases responsible for conversion of glycogen synthase 'a' to glycogen synthase 'b' or inactivate a phosphoprotein phosphatase, thereby preventing the dephosphorylation of glycogen synthase 'b' to glycogen synthase 'a'.

Covalent modification

Covalent modification involves the phosphorylation and dephosphorylation of glycogen synthase catalyzed by protein kinase and PP-1, respectively.

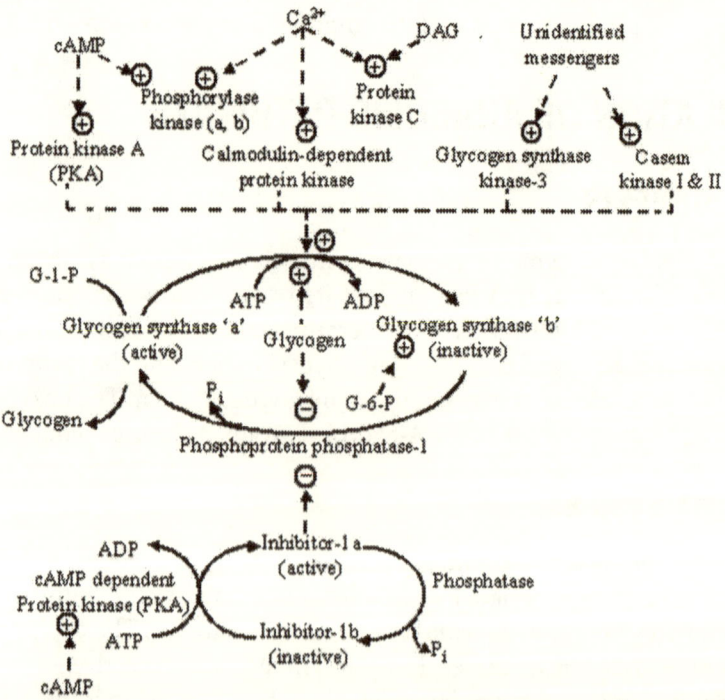

Fig. 3.11: Regulation of glycogen synthase by covalent modification and allosteric influence. (-) inactivation; (+) activation; (− − ▶) allosteric influence.

Protein kinase(s)

Phosphorylation of glycogen synthase 'a' is catalyzed by eight different kinases *in vitro*, which in turn are regulated by several different second messengers of hormonal action, e.g. cAMP, Ca^{2+} and diacylglycerol (DAG) (Fig. 3.11). Each α-subunit of the tetrameric glycogen synthase has nine serine residues at which phosphorylation can take place. Since there are eight kinases capable of phosphorylating each serine residue, there is an amplification process in the phosphorylation of glycogen synthase 'a' and hence its inactivation.

Phosphoprotein phosphatase-1

The dephosphorylation of the 'b'-form of glycogen synthase is catalyzed by a number of different phosphatases. PP-1 is of primary physiological importance and the I-1 protein controls its activity, which in turn is also under the control of PKA. This kinase phosphorylates the I-1 protein and it is this phosphorylated form of the inhibitor (I-1a) that is the active form, which inhibits PP-1 of glycogen synthase thus keeping glycogen synthase inactive. On the other, hand the dephosphorylated form of the inhibitor (I-1b) is produced by the actions of a phosphatase and represents the inactive form of the inhibitor protein. This lifts off the inhibition of PP-1 leading to the activation of glycogen synthase.

Cyclic AMP, therefore, acts in two ways in causing the inhibition of glycogen synthesis: i) phosphorylation of glycogen synthase 'a' through PKA or phosphorylase kinase leading to its inactivation and ii) inhibition of PP-1 through activation of PKA. Conversely, high levels of insulin activate cAMP phosphodiesterase to lower cAMP levels resulting in: i) inhibitor-1 inactivation with the promotion of glycogen synthase dephosphorylation and ii) promotion of the dephosphorylation of inhibitor-1 by the activation of a phosphatase, and consequent activation of glycogen synthase.

GLUCONEOGENESIS

Gluconeogenesis is the biosynthesis of carbohydrates from non-carbohydrate precursors, e.g. lactate, amino acids like alanine, propionate (obtained from FA and amino acid breakdown) and glycerol (obtained from fat catabolism). This occurs in certain mammalian tissues, primarily liver and kidney. It may be taken as a reversal of glycolysis with the only difference being the reversal of the 3 regulatory steps, which are not reversible by the same glycolytic enzymes. This, therefore, requires new enzymatic steps to make reversal possible. These steps are: i) conversion of G-6-P to glucose catalysed by G-6-Pase, ii) conversion of F-1,6-BP to F-6-P by F-1,6-BPase and iii) conversion of pyruvate to PEP in a 2-step process catalysed by pyruvate carboxylase (PC) and PEP carboxykinase (PEPCK) as shown in Fig. 3.12. These enzymes used to reverse glycolysis are primarily involved in the regulation of the pathway.

Inhibition of glycolysis at its regulatory sites or the repression of the synthesis of enzymes involved at these sites (hexokinase/glucokinase, PFK-1 and PK), greatly increases the effectiveness of the opposing gluconeogenic enzymes. Thus, turning on gluconeogenesis is accomplished in a large part by shutting off glycolysis. These regulatory enzymes of gluconeogenesis undergo fine and coarse control mechanisms under hormonal control, which involve the regulation of:

- The supply of FAs and substrates to the liver.
- The enzymes of both the glycolytic and gluconeogenic pathways.

Fine control

Allosteric

Epinephrine and glucagon increase the rate of β-oxidation of FAs in liver by promoting lipolysis in adipose tissue. This results in increased formation of acetyl-CoA and citrate, known allosteric effectors of PC and PFK-1, respectively. Consequently, flux through the glycolytic pathway is curtailed and gluconeogenesis is stimulated . Insulin has opposing effects.

Acetyl-CoA: It is produced from mitochondrial FA oxidation and is a positive allosteric effector of the mitochondrial PC. Increases in the concentration of acetyl-CoA and thus the activation of PC result in a greater rate of synthesis of citrate. The latter is a negative allosteric effector of both PFK-2 and PFK-1, resulting in decreased production of F-2,6-BP and F-1,6-BP, respectively. Hence glycolysis is inhibited and gluconeogenesis promoted.

F-2, 6-BP: It is a positive allosteric effector of PFK-1 and a negative allosteric effector of F-1,6-BPase and, therefore, reduced levels will lead to inactivation of PFK-1 and activation of F-1,6-BPase whereas F-1,6-BP is a positive allosteric effector of PK and, therefore, low levels will inactivate PK. This decreases flux from PEP to pyruvate and increases the effectiveness of the combined effects of PC and PEPCK to convert pyruvate to PEP, thus stimulating gluconeogenesis.

ATP/AMP: An increase in ATP levels with a consequent reduction in AMP levels favours gluconeogenesis by causing the inhibition of PFK-1 and PK and the activation of F-1,6-BPase.

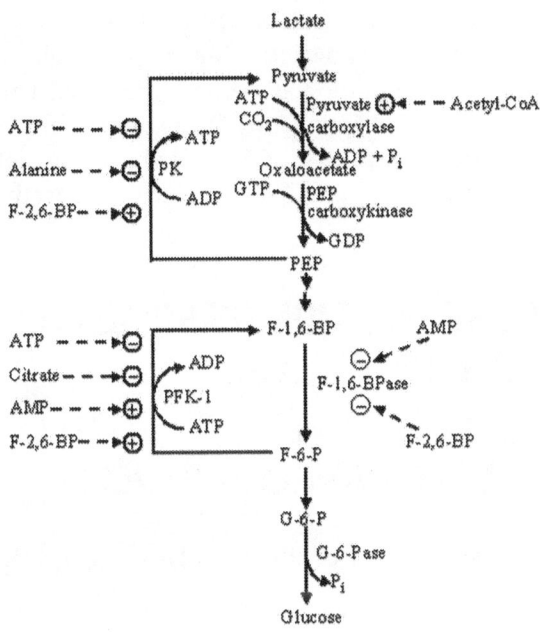

Fig. 3.12: Allosteric regulation of gluconeogenesis. (-) inactivation; (+) activation; (-- ►) allosteric influence

Covalent modification

Glucagon and insulin also regulate gluconeogenesis by influencing the state of phosphorylation of hepatic glycolytic enzymes. First, a high glucagon/insulin ratio activates adenylate cyclase to convert ATP to cAMP, which in turn stimulates phosphorylation of PK from its active dephosphory-lated state to the inactive phosphorylated state. This inactivation of a glycolytic enzyme stimulates gluconeogenesis.

Second, high levels of cAMP cause the phosphorylation of the bifunctional enzyme PFK-2/F-2,6-BPase leading to the inactivation of PFK-2 and the activation of F-2,6-BPase. Thus, F-2,6-BP levels decline and since F-2,6-BP is a positive allosteric effector of PFK-1 low levels will lead to the inactivation PFK-1 and the activation of F-1,6-BPase.

The overall effect of high glucagon/insulin ratio is increased conversion of F-1,6-BP to F-6-P with the corresponding increase in the rate of gluconeogenesis.

Coarse control

A high glucagon/insulin ratio controls the levels of the key gluconeogenic enzymes, PEPCK and G-6-Pase, by activating transcription of the structural genes. A low glucagon/insulin ratio has the opposite effect. Through the inhibition of PEPCK gene transcription, insulin depresses gluconeogenic flux. Glucagon also represses the synthesis of PK and glucokinase, effects that promote flux from pyruvate to PEP and G-6-P to glucose, respectively resulting in the stimulation of gluconeogenesis.

RECIPROCAL REGULATION OF SOME ANABOLIC AND CATABOLIC PATHWAYS

GLYCOGENOLYSIS AND GLYCOGENESIS

The two opposing pathways of glycogenolysis and glycogenesis can be observed in the liver or muscle.

Liver

Glycogenolysis and glycogenesis in the liver are under specialised hormonal controls (Fig. 3.13).

Glucagon

Glucagon is released by α-cells of the pancreas in response to low levels of glucose in the blood (e.g. in a fasted state) or as a result of stress. It helps in the mobilisation of liver glycogen by stimulating glycogenolysis as follows:

a) It interacts with glucagon receptors of liver plasma membrane stimulating adenylate cyclase to produce cAMP, which in turn activates glycogen phosphorylase and inactivates glycogen synthase via PKA.

b) It inhibits the use of glucose via glycolysis at the level of PFK-1 and PK. The net result is the rapid increase in blood glucose levels. There is no hyperglycaemia since less glucagon is released as blood glucose levels rise.

Epinephrine

Epinephrine is released into the blood from the chromaffin cells of the adrenal medulla in response to stress. It mobilises liver glycogen by three different mechanisms as follows:

a) It stimulates glucagon release from the α-cells of the pancreas, which in turn stimulates glycogenolysis.

b) It interacts with β-adrenoceptors of the liver cell plasma membrane to activate adenylate cyclase. The resulting cAMP has the same effect as glucagon.

c) It also interacts with the plasma membrane α-adrenoceptors stimulating the formation of inositol 1,4,5-trisphosphate (IP3) and DAG produced by the action of phospholipase C on phosphatidylinositol 4,5-bisphosphate (PIP2). IP3 releases Ca^{2+} from endoplasmic reticulum (ER), which in turn activates phosphorylase kinase with the consequent activation of glycogen phosphorylase. Ca^{2+}-mediated activation of phosphorylase kinase, calmodulin-dependent protein kinase and protein kinase C as well as the DAG-mediated activation of protein kinase C, lead to the inactivation glycogen synthase.

The net result is increased release of glucose into the bloodstream from glycogen stored in the liver.

Insulin

Insulin is released from the β-cells of the islets of Langerhans of the pancreas in response to high blood glucose levels. It causes increased utilisation of glucose by the liver by the following mechanisms:

a) Insulin binds to its receptors on the plasma membrane and induces autophosphorylation of the receptor. This leads to the activation of receptor tyrosine kinase with subsequent phosphorylation of a variety of extra-receptor proteins including insulin receptor substrates (IRSs) and phospholipase C (PLC).

b) The active phosphorylated PLC hydrolyses glycosylphosphatidyl inositol producing inositol phosphate glycan, which is thought to activate PP-1. This causes dephosphorylation of glycogen synthase, glycogen phosphorylase and phosphorylase kinase with consequent stimulation of glycogenesis and the inhibition of glycogenolysis.

c) Interactions between IRSs and phosphatidylinositol 3-kinase (PI3K) cause the activation of P13K, which then generates PIP3 leading to the activation of protein kinase B and atypical protein kinase C isoforms. These protein kinases activate PP-1, once again stimulating glycogenesis and inhibiting glycogenolysis.

d) These mediators also stimulate cAMP phosphodiesterase activity causing the hydrolysis of cAMP to AMP leading to the inhibition of glycogenolysis and stimulation of glycogenesis.

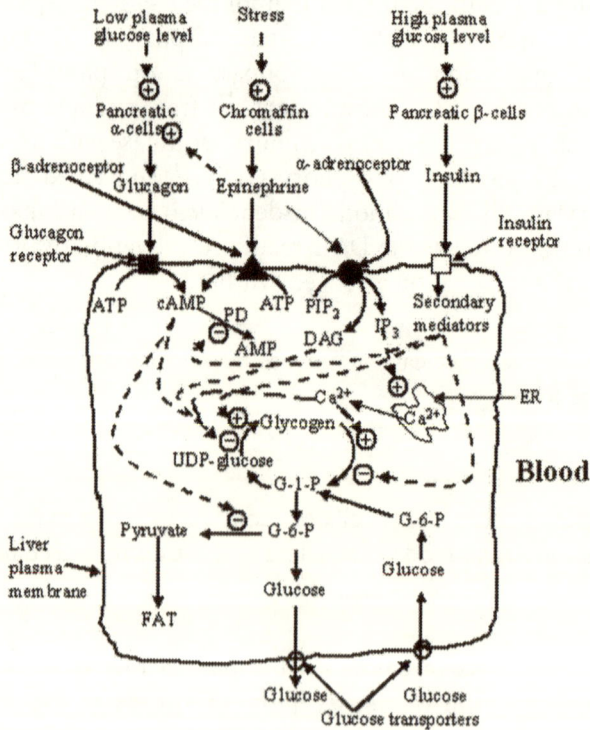

Fig. 3.13: Hormonal regulation of glycogen metabolism in hepatocytes. Actions of epinephrine, glucagon and insulin occur under different conditions ER = endoplasmic reticulum; PD = cAMP phosphodiesterase; (-) inactivation; (+) activation; (--▶) allosteric influence.

Muscle

Regulation of glycogen metabolism in the muscle is under hormonal and neural control generally in response to stressful conditions or increased muscular activity (Fig. 3.14).

Epinephrine

Epinephrine is released in response to stress and stimulates glycogen degradation. It acts on β-adrenoceptors of plasma membrane activating adenylate cyclase and increasing the production of cAMP. The consequent activation of glycogen phosphorylase and inactivation of glycogen synthase lead to glycogen breakdown to G-1-P and then to G-6-P does not end up in glucose production since muscle does not possess G-6-Pase. Thus, more G-6-P is made available for glycolysis in the muscle. The ATP produced can be used to meet the metabolic demands imposed on the skeletal muscle by the stress that triggered epinephrine release.

Neural control

Stimulation of acetylcholine receptors at the plasma membrane as a result of nervous impulse via acetylcholine leads to the depolarisation of the sarcoplasmic membrane with the release of Ca^{2+} into the cytoplasm, which triggers muscle contraction. Reaccumulation of Ca^{2+} into the sarcoplasmic reticulum leads to muscle relaxation. With increasing muscle contraction there is a greater need for ATP, which is met by the indirect activation of glycogen phosphorylase by Ca^{2+} and its inactivation of glycogen synthase. Thus, more glycogen is converted to G-6-P for more ATP to be produced by glycolysis to meet the greater energy requirements of a contracting muscle. There appears to be a synergistic effect of neural signals and epinephrine on glycogen degradation in muscle.

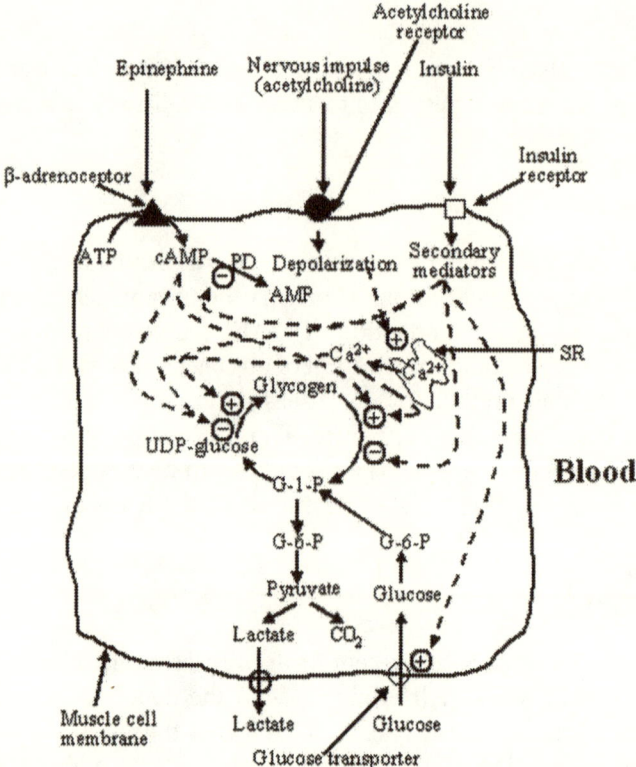

Fig. 3.14: Hormonal and neural regulation of glycogen metabolism in muscle. SR = sarcoplasmic reticulum; PD = cAMP phosphodiesterase; (-) inhibition; (+) activation; (- - ▶) allosteric influence.

Insulin

The action of insulin in the muscle appears to be similar in most respects to that found in the liver. It increases glucose utilisation rates through secondary mediators by:

a) Promoting glycogenesis and inhibiting glycogenolysis by stimulating dephosphorylation of both glycogen phosphorylase and glycogen synthase through the inhibition of protein kinases and activation of phosphoprotein phosphatases.

b) Stimulation of glucose transport into muscle cells by signalling an increase in the number of functional glucose transporters associated with the plasma membrane. This incidentally does not normally occur in liver cells because the glucose transporters of liver plasma membrane are not responsive to insulin stimulation.

GLYCOLYSIS AND GLUCONEOGENESIS

Glycolysis catabolizes glucose while gluconeogenesis synthesises glucose, thus they must be controlled in a reciprocal fashion, i.e. catabolism of glucose requires that glucose synthesis is curtailed and vice versa in the same tissue. These two opposing pathways that take place predominantly in the cytosol can operate at the same time in two different tissues, e.g. gluconeogenesis in the liver and glycolysis in the muscle.

Low tissue energy levels lead to the activation of the glycolytic pathway while high tissue energy levels lead to the activation of the gluconeogenic pathway.

Further reading

Chayen, J., Howat, D.W., and Bitensky, L. (1986). Cellular Biochemistry of Glucose-6-phosphate and 6-Phosphogluconate dehydrogenase Activities. *Cell Biochem. Funct.* 4:249-253

Devlin T.M. (1992). Textbook of Biochemistry with Clinical Correlations. Wiley—Liss Inc., New York

Hue, L., and Rider, M.H. (1987). Role of Fructose-2,6-bisphosphate in the Control of Glycolysis in Mammalian Tissues. *Biochem. J.* 245:313-324

Nelson, D.L., and Cox, M.M. (2000). Lehninger Principles of Biochemistry. Worth Publishers, New York

Oliver, C.J., and Shenolikar, S. (1998). Physiological Importance of Protein Phosphatase Inhibitors. *Front. Biosci.* 3:D961-972

Phillips, D., Blake, C.C.F., and Watson, H.C. (eds). (1981). The Enzymes of Glycolysis: Structure, Activity and Evolution. *Philos. Trans. R. Soc. Lond. [Biol.]* 293:1-214

CHAPTER FOUR

LIPID METABOLISM

Lipid metabolism involves the hydrolysis of TAGs and the oxidation (degradation) of released FAs for the provision of energy. It also involves the synthesis of TAGs, phospholipids and cholesterol, which are utilised in processes such as membrane biogenesis or steroid biosynthesis or for storage.

REGULATION OF TRIACYLGLYCEROL DEGRADATION

The major source of energy in lipids is FAs. They are stored in adipocytes esterified as TAGs. The release of fatty acids from the TAGs through lipolysis for provision of energy by β-oxidation is under hormonal control.

Lipolysis

Lipolysis involves the hydrolysis of TAG in adipose tissue, which occurs at high blood levels of epinephrine and or glucagon and low levels of insulin under stressful or starved states, respectively. Epinephrine and glucagon increase the rate of lipolysis while insulin has the opposite effect (Fig. 4.1).

Insulin

The inhibition of lipolysis in adipocytes and myocytes by insulin occurs through the activation of a family of cAMP phosphodiesterases, which catalyses the breakdown of cAMP to AMP. This reverses the activation of hormone sensitive lipase (HSL), which is responsible for TAG hydrolysis to free fatty acids (FFAs) and glycerol. The AMP may also allosterically inhibit active HSL, thus preventing the hydrolysis of TAG.

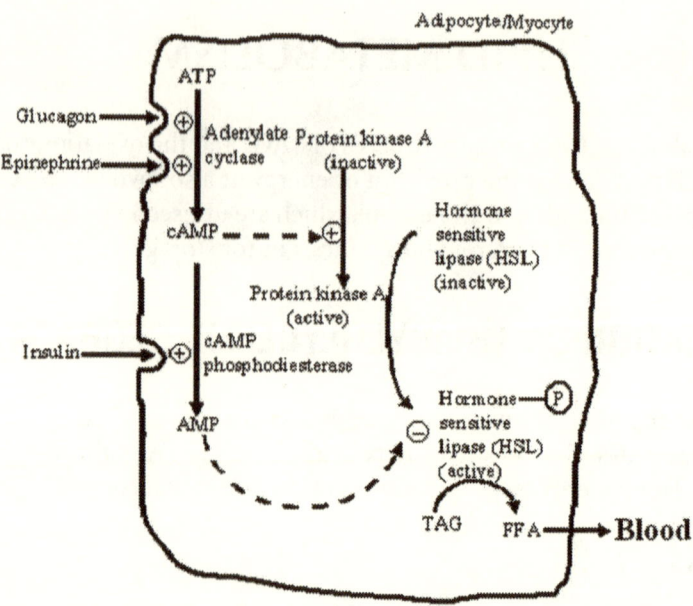

Fig. 4.1: Hormonal control of triacylglycerol (TAG) degradation (lipolysis) in adipocytes and myocytes. (-) inhibition; (+) activation; (- - ▶) allosteric influence

Glucagon/Epinephrine

Glucagon and epinephrine act by binding to glucagon and β-adrenergic receptors of adipocytes, respectively, activating adenylate cyclase, which cataly-ses the formation of cAMP. High cAMP levels activate protein kinase A (PKA) responsible for the phosphorylation and subsequent activation of HSL, which in turn catalyses the hydrolysis of TAG to FFA and monoacylglycerols (MAG). The MAG can be converted further to glycerol and FFA by either HSL or MAG lipase. The glycerol and FFAs enter the bloodstream via the plasma membrane of adipocytes. Glycerol goes to the liver and is converted to glucose through gluconeogenesis since adipose tissue lacks glycerol kinase, a key enzyme required in the conversion of glycerol to glycerol 3-P. FFA are bound to serum albumin and carried to liver and other tissues like heart and skeletal muscle and oxidized in the mitochondria to release energy.

β-Oxidation

Mammalian catabolism of FAs in the mitochondrion, called **β-oxidation**, is also under hormonal control. Epinephrine and glucagon are the principal hormones that control FA catabolism in the liver (Fig. 4.2).

When glycogen stores are depleted there is a switch from the use of glucose to FA by most tissues as energy source. The oxidation of FAs is regulated intracellularly in the liver by the control of the formation of acylcarnitine and thus the control of transport of FAs into the mitochondria by the carnitine shuttle system; carnitine acyltransferase I (CAT I). In the liver CAT I is allosterically inhibited by malonyl-CoA, an intermediary product of FA synthesis, whose production is promoted by insulin and inhibited by glucagon and epinephrine.

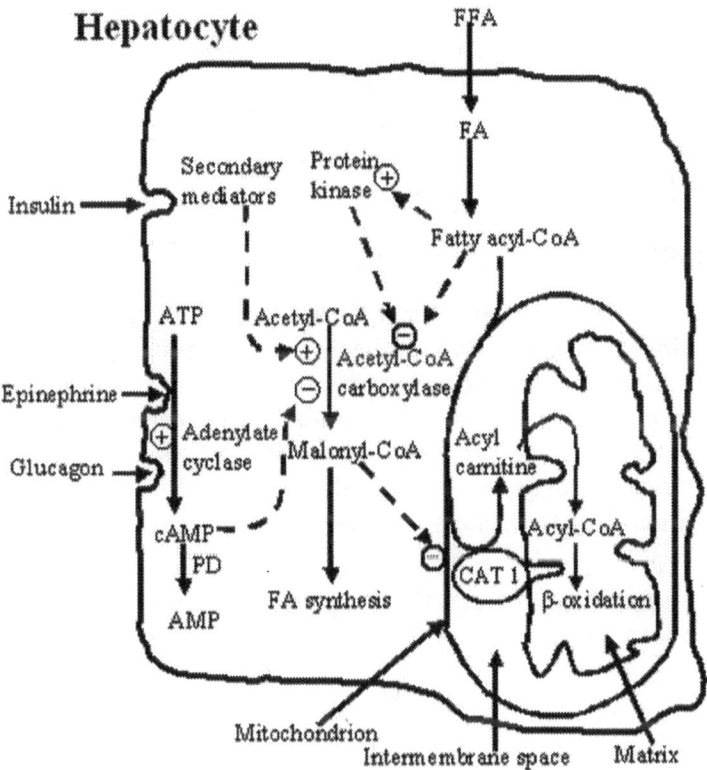

Fig. 4.2: Hormonal control of fatty acid metabolism in liver. (-) inhibition; (+) activation;(- - ▶) allosteric influence; PD = cAMP phosphodiesterase; CAT I = Carnitine acyltransferase I

Glucagon/Epinephrine

Glucagon, and to some extent epinephrine, stimulates the phosphorylation and inactivation of the acetyl-CoA carboxylase (ACC), through the activation of liver plasma membrane adenylate cyclase, depressing malonyl-CoA formation and thus the rate of FA synthesis. The ACC is further regulated by fatty acyl-CoA, which increases the activity of a protein kinase that catalyses the phosphorylation and inhibition of ACC. Fatty acyl-CoA at higher concentrations is also an allosteric inhibitor of ACC. The net results are low levels of malonyl-CoA, causing a lifting of the inhibition of CAT I, and allowing for the easy transport of FA as fatty acylcarnitine into the mitochondrial matrix for β-oxidation.

Insulin

Insulin opposes the lipolytic actions of glucagon and epinephrine. It stimulates the formation of secondary mediators, which in turn stimulate ACC causing an increase in the rate of malonyl-CoA formation. Malonyl-CoA is an allosteric inhibitor of CAT I, thus transport of FA into the mitochondrial matrix for β-oxidation is curtailed.

Ketogenesis

Acetyl-CoA formed in the liver through β-oxidation of FAs will normally enter the citric acid cycle (CAC) for the production of energy. However, under conditions where the rate of production of acetyl-CoA exceeds the capacity of the CAC to oxidise it, the excess acetyl-CoA molecules combine to form ketone bodies (β-hydroxybutyrate and acetoacetate) in a process called **ketogenesis.**

In mammals, ketogenesis occurs in the mitochondrial matrix of the liver. The formation of acetoacetate is a three-step process:

1) Formation of acetoacetyl-CoA from two molecules of acetyl-CoA catalysed by a thiolase.

$$2 \text{ Acetyl-CoA} \longrightarrow \text{acetoacetyl-CoA} + \text{CoASH}$$

2) Another molecule of acetyl-CoA reacts with acetoacetyl-CoA to form 3-hydroxy-3-methylglutaryl-CoA (HMG-CoA) in a reaction catalysed by HMG-CoA synthase.

Acetoacetyl-CoA + acetyl-CoA $\longrightarrow$ HMG-CoA

3) HMG-CoA is cleaved by HMG-CoA lyase to acetoacetate and acetyl-CoA.

HMG-CoA $\longrightarrow$ acetoacetate + acetyl-CoA

Acetoacetate can be converted to β-hydroxybutyrate by an NADH-dependent reduction catalysed by β-hydroxybutyrate dehydrogenase.

Acetoacetate + NADH + H$^+$ $\longrightarrow$ β-hydroxybutyrate + NAD$^+$

Both β-hydroxybutyrate and acetoacetate can be transported across the mitochondrial and plasma membranes of liver cells into the bloodstream and to peripheral tissues to be used as fuel. Small amounts of acetoacetate are decarboxylated to acetone in the bloodstream.

Conditions for excess ketone body formation

Under conditions of severe starvation or untreated diabetes mellitus (DM) there is excessive production of ketone bodies. During starvation, formation of glucose by gluconeogenesis leads to the depletion of CAC intermediates like oxaloacetate, thus diverting acetyl-CoA to ketone body formation. In untreated DM, where there is insufficient uptake of glucose by extrahepatic tissues from the blood to use as fuel or for conversion to fat for storage, the formation of malonyl-CoA is curtailed and thus the inhibition of CAT I is lifted. This leads to the transfer of FAs into the mitochondrial matrix for oxidation to acetyl-CoA, which cannot be fully utilised by the CAC because of the use of cycle intermediates like oxaloacetate in gluconeogenesis. The resultant accumulation of acetyl-CoA may accelerate the formation of ketone bodies beyond the capacity of extrahepatic tissues to oxidize, leading to the condition called ketoacidosis.

Ketone bodies as fuels

Extrahepatic tissues can utilize ketone bodies for the production of energy for biosynthetic processes. β-Hydroxybutyrate can be oxidized to acetoacetate by β-hydroxybutyrate dehydrogenase (1). Acetoacetate is activated to acetoacetyl-CoA by the transfer of CoA from succinyl-CoA catalysed by β-ketoacyl-CoA transferase (2). The acetoacetyl-CoA is then cleaved by a thiolase (3) to yield two molecules of acetyl-CoA, which enter the CAC.

$$\beta\text{-Hydroxybutyrate} + NAD^+ \underset{}{\overset{1}{\rightleftharpoons}} \text{acetoacetate} + NADH + H^+$$

$$\text{Acetoacetate} + \text{succinyl-CoA} \underset{}{\overset{2}{\rightleftharpoons}} \text{acetoacetyl-CoA} + \text{succinate}$$

$$\text{Acetoacetyl-CoA} + \text{CoASH} \underset{}{\overset{3}{\rightleftharpoons}} 2 \text{ acetyl-CoA}$$

REGULATION OF LIPID SYNTHESIS

Although the term lipid refers to among others; TAGs, phosphoglycerides, and cholesterol, this section is devoted to the regulation of TAG and cholesterol synthesis.

Fatty Acid/TAG Synthesis

In the fed state, insulin levels rise in response to high blood glucose levels and stimulate the biosynthesis of FA/TAG. Insulin acts on both adipocytes and hepatocytes in the regulation of FA and TAG synthesis. Binding of insulin to adipocytes causes the activation of cAMP phosphodiesterase leading to the conversion of cAMP to AMP. Low levels of cAMP prevent the activation of PKA responsible for the activation of HSL while high levels of AMP cause the allosteric inhibition of active HSL. The net result is the inhibition of lipolysis (Fig. 4.1).

Insulin also stimulates ACC in hepatocytes through secondary mediators, resulting in an increase in the rate of malonyl-CoA formation and hence the rate of synthesis of FAs and TAG. Malonyl-CoA allosterically inhibits CAT I, preventing the entry of FAs into the mitochondrial matrix for β-oxidation (Fig. 4.2). *In vitro* activation of ACC by citrate has led to the suggestion of a feed-forward mechanism for the activation of FA synthesis since cytosolic citrate is a precursor of acetyl-CoA used in FA synthesis. Epinephrine activates 5' AMP kinase (AMPK) leading to the phosphorylation of ACC and its subsequent inhibition.

Cholesterol Biosynthesis

Cholesterol biosynthesis occurs in the cytoplasm and is regulated at the level of HMG-CoA reductase, an integral membrane protein of the endoplasmic reticulum, and at the low-density lipoprotein (LDL) receptor of the plasma membrane. It involves both fine and coarse controls (Fig. 4.3).

Covalent modification

The hormones insulin and glucagon regulate the activity of HMG-CoA reductase by covalent modification of the enzyme. In a fasted state, glucagon stimulates the conversion of active HMG-CoA reductase by phosphorylation to the inactive enzyme, thereby inhibiting cholesterol biosynthesis. Insulin in a fed state, on the other hand, stimulates the dephosphorylation of the inactive HMG-CoA reductase to the active enzyme thus stimulating cholesterol biosynthesis.

Coarse control

- High cholesterol levels inhibit the uptake of cholesterol-rich LDL at the plasma membrane by repressing the transcription of the gene for the LDL receptor.

- High cholesterol levels can also repress the transcription of genes that encode HMG-CoA reductase thereby inhibiting its activity.

- Farnesylated protein products from farnesyl pyrophosphate, an intermediate in cholesterol biosynthesis, appear to suppress the translation of mRNA encoding the HMG-CoA reductase.

- High cholesterol and farnesylated protein levels increase the activity of a protease, which then attacks HMG-CoA reductase degrading it and thus reducing its activity.

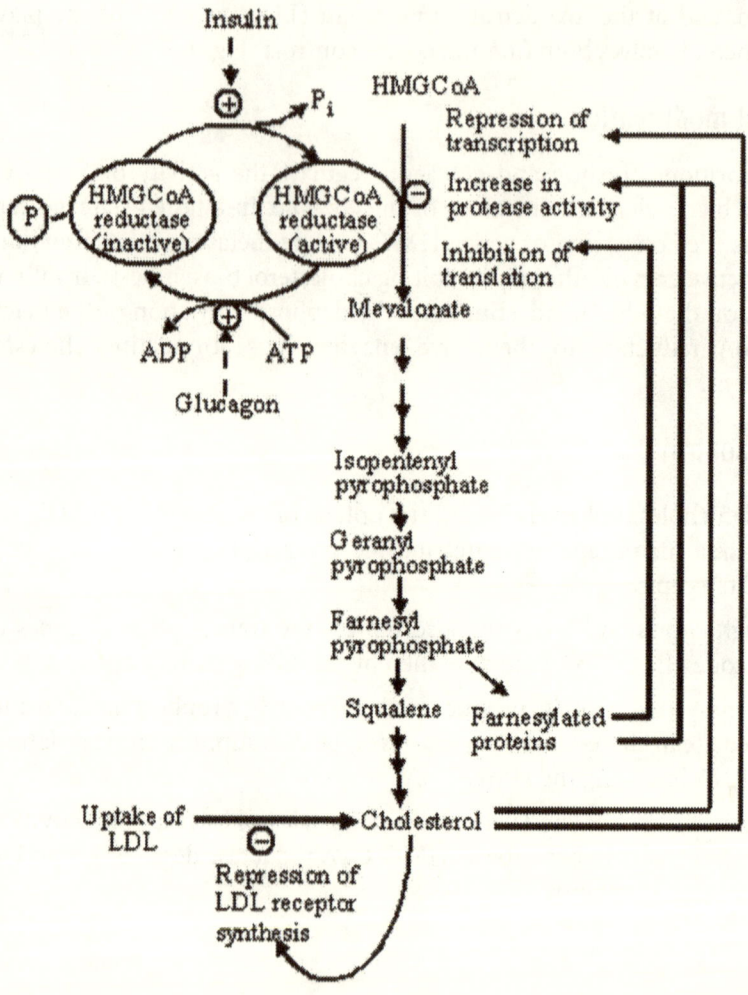

Fig. 4.3: Regulation of cholesterol biosynthesis. (-) inhibition; (+) stimulation .

Further reading

Fernandez-Figares, I., Shannon, A.E., Wray-Cahen, D., and Caperna, T.J. (2004). The Role of Insulin, Glucagon, Dexamethasone, and Leptin in the Regulation of Ketogenesis and Glycogen Storage in Primary Cultures of Porcine Hepatocytes prepared from 60 kg pigs. *Domest. Anim. Endocrinol.* 27:125-140.

Mabrouk, G.M., Helmy, I.M., Thampy, K.G., and Wakil, S.J. (1990). Acute Hormonal Control of Acetyl-CoA Carboxylase: The Roles of Insulin, Glucagon, and Epinephrine. *J. Biol. Chem.* 265:6330-6338

Nelson, D.L. and Cox, M. (2000). Lehninger Principles of Biochemistry. Worth Publishers, New York

Numa, S. (ed). (1984), Fatty Acid Metabolism and Its Regulation. Elsevier Science Publishing Co. Inc., New York

Sul, H.S., Smas, C.M., Wang, D., and Chen, L. (1998). Regulation of Fat Synthesis and Adipose Differentiation. *Prog. Nucleic Acid Res. Mol. Biol.* 60:317-345

Vance, D.E., and Vance, J.E. (eds.). (1996). Biochemistry of Lipids, Lipoproteins, and Membranes. Elsevier Science Publishing Co. Inc., New York

Panda, T., and Devi, V.A. (2004). Regulation and Degradation HMG-CoA Reductase. *Appl. Microbiol. Biotechnol.* 66:143-152

CHAPTER FIVE

NITROGEN METABOLISM

Biosynthetic pathways that lead to production of amino acids and nucleotides require nitrogen. It is therefore important to understand how nitrogen from the environment is incorporated into biological molecules.

GENERAL NITROGEN METABOLISM

Atmospheric nitrogen is fixed in prokaryotes by enzymes of the nitrogenase complex. The first important product of nitrogen fixation is ammonia (NH_3), which can be used by other organisms either directly or indirectly after its conversion to other soluble compounds like nitrite, nitrates and amino acids.

In an aqueous environment ammonia exists in the form of NH_4^+ and is assimilated into amino acids and then into other nitrogen-containing biomolecules like purines and pyrimidines. The amino acids, glutamine and glutamate provide a critical point of entry of nitrogen into the biosynthetic pathways of other nitrogenous compounds. The biosynthetic pathways of glutamate and glutamine occur in most organisms. The most important pathway for the assimilation of NH_4^+ into glutamate requires two reactions (Fig. 5.1).

a) The conversion of glutamate and NH_4^+ to glutamine, by glutamine synthetase. The enzyme also has a central role in amino acid metabolism in mammals, in converting toxic NH_4^+ into glutamine.

b) In bacteria and plants, glutamate is produced from glutamine in a reaction catalysed by glutamate synthase involving the reductive amination of α-ketoglutarate with glutamine as nitrogen donor. Since glutamate synthase is not present in animals, glutamate is obtained from the transamination of α-ketoglutarate during amino acid catabolism. It can also be obtained to a lesser extent by a reaction between α-ketoglutarate and NH_4^+ catalysed by glutamate dehydrogenase present in all organisms using NADPH or NADH as reducing equivalents.

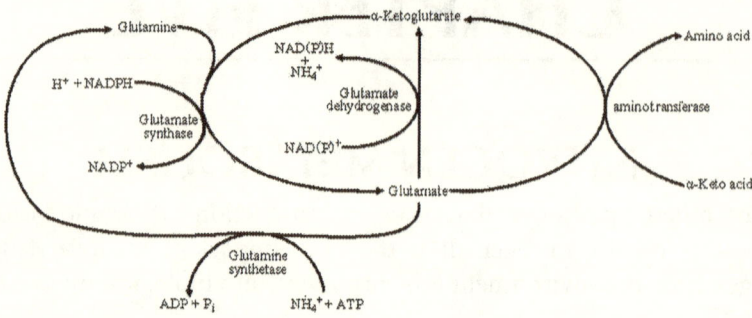

Fig. 5.1: Actions of glutamine synthetase, glutamate synthase and glutamate dehydrogenase in the metabolism of some amino acids

Regulation of Glutamate Dehydrogenase

In mammalian hepatocytes, glutamate is transported from the cytosol into the mitochondria, where it undergoes oxidative deamination to α-ketoglutarate catalysed by L-glutamate dehydrogenase, which is present in the mitochondrial matrix. L-glutamate dehydrogenase also catalyses a highly efficient route for the incorporation of NH_3 into the amino acids in plants and animals by the reductive amination of α-ketoglutarate.

Allosteric control

Glutamate dehydrogenase is an allosteric enzyme with six identical subunits whose activity is influenced by an array of allosteric modulators. ADP is a positive allosteric effector while GTP is a negative allosteric effector (Fig. 5.2).

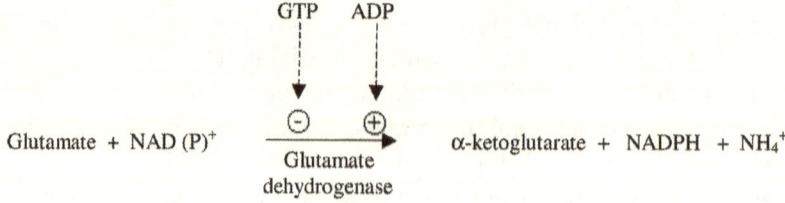

Fig. 5.2: Allosteric regulation of glutamate dehydrogenase activity. (-) inactivation; (+) activation;($--\blacktriangleright$) allosteric influence.

Regulation of Glutamine Synthetase

The activity of glutamine synthetase is regulated in all organisms since it is the entry point for reduced nitrogen in the biosynthesis of amino acids and other nitrogen-containing biomolecules. In bacteria this regulation is unusually complex and involves fine and coarse controls.

Fine control

Allosteric

The enzyme in *E. coli* consists of 12 subunits each of M_r 51,600. Each subunit has a catalytic site and a regulatory site. The regulatory site has allosteric sites for 9 inhibitors namely AMP, carbamoyl phosphate, CTP, histidine (his), tryptophan (trp), glucosamine-6-phosphate, alanine (ala), serine (ser) and glycine (gly). Six of the inhibitors contain a nitrogen atom obtained directly from the amide nitrogen of glutamine. The other three inhibitors ala, ser and gly have their nitrogen atom obtained indirectly from glutamate. In mammals, glutamine synthetase of liver and brain contain 8 identical subunits, which exist as tetramers. Carbamoyl phosphate, ser, ala and gly are inhibitors of the mammalian enzyme while α-ketoglutarate is an activator.

Each inhibitor on its own produces a partial inhibition of glutamine synthetase. The degree of inhibition increases as more and more inhibitors are bound, a process called **cumulative feedback inhibition** (Fig. 5.3).
Table 5.1 shows that when saturating concentrations of 4 of the 9 inhibitors were individually incubated with the enzyme *in vitro*, varying degrees of inhibition (13–41%) were obtained but when all 4 inhibitors were combined 63% inhibition was obtained. When all 9 inhibitors were combined the inhibition of the enzyme was almost complete.

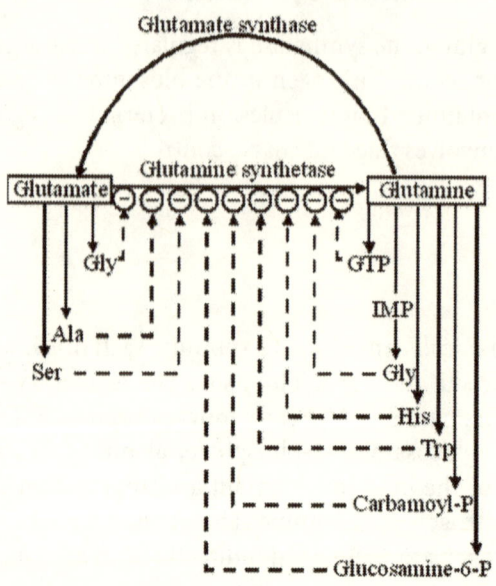

Fig. 5.3: Allosteric inhibition of glutamine synthetase activity in *E. coli*. (-) inhibition; (--►) allosteric influence.

Table 5.1: Effects of allosteric inhibitors on glutamine synthetase activity

Saturating inhibitor	% Inhibition
Trp	16
CTP	14
Carbamoyl phosphate	13
AMP	41
All 4 above	63
All 9 inhibitors	98

Covalent modification

Covalent modification in *E. coli* occurs as part of a regulatory cascade that includes the sequential addition of AMP moieties to tyrosine (tyr) 397 of glutamine synthetase (adenylylation), present on each of the 12 subunits, leading to inactivation of the enzyme. Removal of the AMP moieties (deadenylylation) results in the activation of the enzyme. Both adenylylation and deadenylylation of glutamine synthetase are catalysed by the same enzyme adenylyltransferase (ATase) as in Fig. 5.4.

The action of ATase that predominates is determined by another protein called P_{II}, which can form a complex with the ATase. The P_{II} itself is subject to covalent modification by the reversible transfer of a UMP group to the phenolic hydroxyl of a specific tyr residue of P_{II}, catalysed by uridylyltransferase, leading to activation of glutamine synthetase. Uridylylation is stimulated by α-ketoglutarate and inhibited by glutamine. The removal of UMP from P_{II} is catalysed by a uridylyl removing enzyme culminating in the deactivation of glutamine synthetase (Fig. 5.4).

Mammalian glutamine synthetase, however, does not undergo covalent modification.

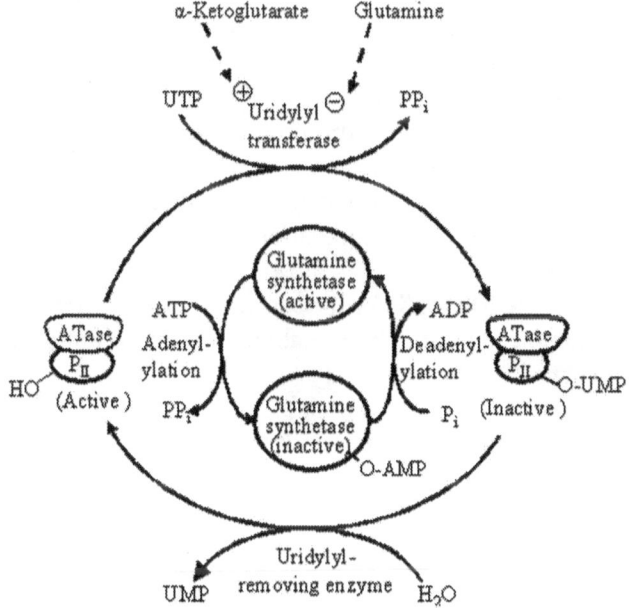

Fig. 5.4: **Regulation of glutamine synthetase of *E. coli* by covalent modification.** (-) inactivation; (+) activation;(- - ▶) allosteric influence.

Coarse control

Under conditions in which *E. coli* is grown in nitrogen-limited medium, higher levels of glutamine synthetase are produced through increased transcription of the gene encoding for it and subsequent translation of the corresponding mRNA. The synthesis may be increased more than a hundred fold.

AMINO ACID METABOLISM

Transamination

Enzymes called transaminases or aminotransferases catalyse transamination reactions, which involve the transfer of the amino group of for example glutamate to an α-keto acid with the formation of the corresponding amino acid and the keto derivative of glutamate, namely α-ketoglutarate. In the presence of the requisite α-keto acids, aminotransferases of mammalian cells synthesize most amino acids contained in proteins.

Since transamination reactions have equilibrium constants close to unity. The direction in which a particular transamination reaction will proceed is controlled to a large extent by the intracellular concentrations of substrates and products. Thus, it can be used for both synthesis and degradation of amino acids.

With regard to synthesis, because amino acids in a cell are rarely present in proportions needed to synthesize a specific protein of that cell, transamination plays an important role in bringing the amino acid composition into line with the protein requirements of an organism. During degradation, an aminotransferase works together with glutamate dehydrogenase in a process called transdeamination. For example, the degradation of excess alanine results in the formation of NADH and pyruvate, as shown in the net equation below, which are then utilised in the ETC and CAC, respectively for the provision of energy. Thus, transamination can lead to the catabolism of excess amino acids for the generation of energy.

$$\text{Alanine} + \alpha\text{-ketoglutarate} \xrightleftharpoons{\text{aminotransferase}} \text{pyruvate} + \text{glutamate}$$

$$\text{Glutamate} + NAD^+ + H_2O \xrightleftharpoons{\text{glu dehydrogenase}} \alpha\text{-ketoglutarate} + NADH + NH_4^+$$

$$\text{Net}\quad \text{Alanine} + NAD^+ + H_2O \rightleftharpoons \text{pyruvate} + NADH + NH_4^+$$

AMINO ACID BIOSYNTHESIS

Regulation of Amino Acid Biosynthesis

Amino acid synthesis is generally regulated through feedback inhibition of the first reaction, in a sequence of reactions, by the end product of the pathway. The first reaction is usually irreversible and is catalysed by an allosteric enzyme, e.g. synthesis of isoleucine (ile), valine (val) and leucine (leu) in *E. coli* (Fig. 5.5).

In bacteria, each of these amino acids controls its own synthesis by feedback inhibition of a different enzyme. Ile is an allosteric inhibitor of threonine dehydratase, the enzyme that catalyses the first step in the biosynthesis of ile from threonine (thr). The first step in the synthesis of val from pyruvate, catalysed by acetolactate synthase, is allosterically inhibited by val whereas the first step in the biosynthesis of leu from α-ketoisovalerate, catalysed by α-isopropylmalate synthase, is inhibited by leu.

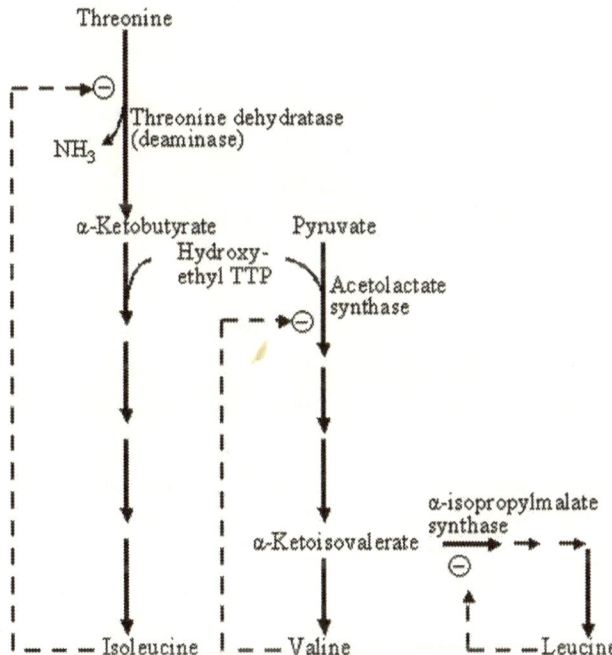

Fig. 5.5: Allosteric regulation of isoleucine, valine and leucine biosynthesis in *E. coli*.
(-) inhibition; (- - ►) allosteric influence.

Coordination of synthesis of Lysine, Methionine, Threonine and Isoluecine

During protein synthesis, twenty amino acids must be available in the right proportions. Some cells have, therefore, developed ways to control both the rate of synthesis and coordination of the formation of each amino acid. This kind of coordination is well developed in fast growing bacterial cells, for the synthesis of lysine (lys), methionine (met), thr and ile from aspartate in *E. coli* (Fig. 5.6).

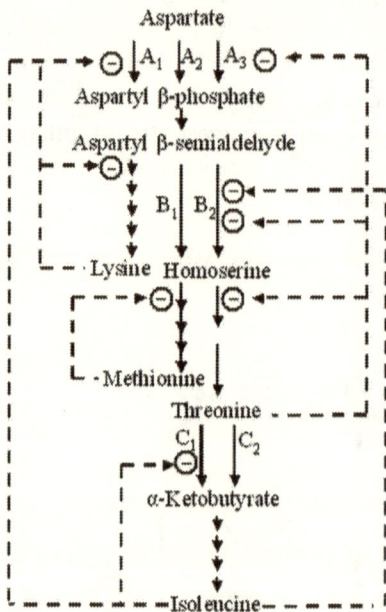

Fig. 5.6: **Coordinating regulatory mechanisms in the synthesis of four aspartate-derived amino acids in *E. coli*.** (-) inhibition; (- - ►) allosteric influence.

The steps from aspartate (asp) to aspartyl β-phosphate is catalysed by three isozymes of aspartokinase designated A_1, A_2 and A_3, with A_1 and A_3 being independently controlled by different allosteric modulators. This prevents one end product from shutting down key steps in a pathway when other products of the same pathway are required. Isozyme A_1 is allosterically inhibited by two different modulators, lys and ile, which show a cumulative feedback inhibition.

The conversion of aspartyl β-semialdehyde to homoserine and thr to α-ketobutyrate are catalyzed by two independently controlled isozymes B_1 and

B_2 of homoserine dehydrogenase and C_1 and C_2 of threonine dehydratase, respectively. Isozymes B_2 and C_1 are under allosteric control while A_2, B_1 and C_2 are under coarse control. Synthesis of A_2 and B_1 is repressed when met levels are high while that of C_2 is repressed when ile levels are high. On the other hand, the reactions from asp to ile show multiple overlapping feedback inhibitions. Ile inhibits the conversion of thr to α-ketobutyrate and thr inhibits its own synthesis at three different points from: (i) homoserine to thr, (ii) aspartyl β-semialdehyde to homoserine and (iii) asp to aspartyl β-phosphate. This mechanism of regulation is referred to as **sequential feedback inhibition.**

AMINO ACID CATABOLISM

The catabolism of amino acids leads to formation of NH_4^+ and carbon skeletons that enter the CAC. The NH_4^+ can be used in the synthesis of amino acids, nucleotides and biological amines or converted to carbamoyl phosphate, which enters the urea cycle and is eventually excreted as urea.

Regulation of Urea Cycle

The urea cycle is regulated by both allosteric and coarse control mechanisms.

Allosteric control

Allosteric control involves at least one key enzyme, carbamoyl phosphate synthetase I, which adjusts the flux through the urea cycle. It is allosterically activated by N-acetylglutamate, which is synthesised from acetyl-CoA and glutamate in the presence of N-acetylglutamate synthetase.

Coarse control

The rates of synthesis of four urea cycle enzymes (ornithine transcarbamoylase, argininosuccinate synthetase, argininosuccinate lyase and arginase), and carbamoyl phosphate synthetase I in the liver are increased in starving animals and those on high protein diets. Animals on protein free diets produce lower levels of urea cycle enzymes, since there is less protein available for degradation.

PROTEIN METABOLISM

Numerous hormones and peptide growth factors are involved in the synthesis and degradation of proteins (e.g. insulin, glucagon, steroid and thyroid hormones). Insulin stimulates the entry of amino acids into cells for protein synthesis and inhibits proteolysis. Recent evidence has shown that insulin activates several signalling cascades including the PI3 kinase pathway. This results in alterations in the phosphorylation states of several eukaryotic initiation and elongation factors (eIFs and eEFs).

NUCLEOTIDE METABOLISM

Nucleotides serve as precursors of nucleic acid synthesis, critical elements in energy metabolism (e.g. ATP, GTP, GDP, ADP), carriers of activated metabolites for biosynthesis (nucleoside diphosphate sugars), structural moieties of coenzymes (e.g. NAD^+, NADH, NADPH, $NADP^+$), and metabolic regulators and signal molecules (e.g. cAMP).

Purines and pyrimidine nucleotides can be synthesised through *de novo* or salvage pathways. *De novo* synthesis involves synthesis from low molecular weight precursors and salvaging involves the utilisation of preformed purine and pyrimidine compounds that would otherwise be lost to biodegradation.

Regulation of Nucleotide Biosynthesis

The control of nucleotide biosynthesis occurs in both *de novo* synthesis and salvage pathways.

Purine nucleotides

De novo synthesis

Purine nucleotide biosynthesis may be regulated in cells by feedback inhibition since several enzymes that catalyse steps in the biosynthesis of purine nucleotides exhibit allosteric kinetic behaviour *in vitro*. The overall rate of purine nucleotide synthesis and the relative rates of formation of the two end products, AMP and GMP, are regulated by three major feedback mechanisms in *E. coli* (Fig. 5.7).

The first mechanism operates at the first step (1) in the biosynthetic pathway involving the conversion of ribose 5-P to 5-phosphoribosyl-1-pyrophosphate (PRPP), catalysed by PRPP synthase. AMP, GMP and IMP each allosterically inhibits this enzyme. The second mechanism involves the committed step (2) in the biosynthesis of purine nucleotides (the second step in the biosynthetic process), the conversion of PRPP to 5-phosphoribosylamine, catalysed by glutamine-PRPP amidotransferase involving the transfer of amino (NH_2) groups from glutamine to PRPP. This enzyme is also allosterically inhibited by each of the end products IMP, AMP and GMP.

The third and final mechanism is at the branch point of IMP to GMP and AMP (3). Adenylosuccinate synthetase catalyses the formation of AMP from IMP through adenylosuccinate while IMP dehydrogenase catalyses the conversion of IMP to XMP and then to GMP. AMP is a competitive inhibitor of adenylosuccinate synthetase while GMP and XMP are competitive inhibitors of IMP dehydrogenase. AMP can be converted to ATP in a two-step process and the ATP formed is the energy source in the conversion of XMP to GMP. The GMP can also be converted to GTP, in a two-step process, which is the energy source for the conversion of IMP to adenylosuccinate.

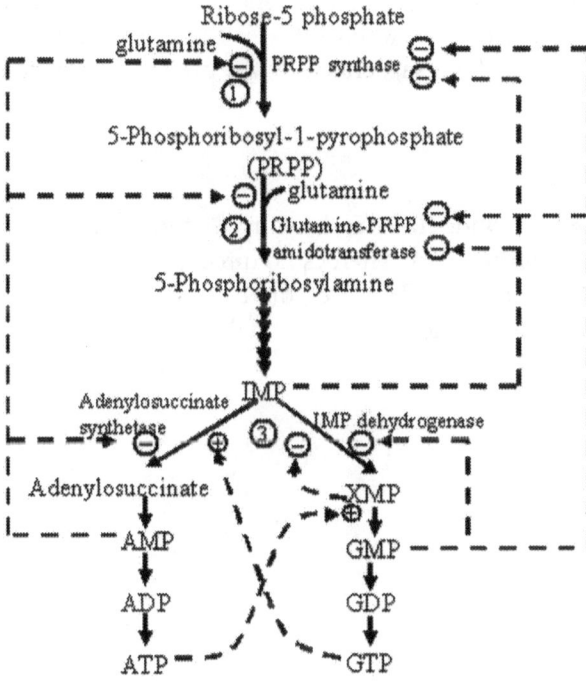

Fig. 5.7: **Regulation of purine nucleotide biosynthesis in** *E. coli.* (-) inactivation; (+) activation; (- - ▶) allosteric influence.

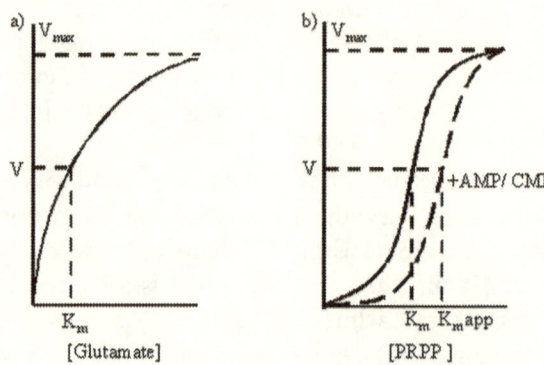

Fig. 5.8: Kinetics of PRPP amidotransferase using a) glutamine and b) PRPP as substrates. The former produces a hyperbolic curve while the latter produces a sigmoidal one. AMP or CMP increases the K_m of the enzyme for PRPP ($K_m$app).

The enzyme PRPP amidotransferase shows hyperbolic kinetics with glutamine as substrate. However, when PRPP is the substrate, it shows sigmoidal kinetics, which is further exaggerated by the presence of nucleotides AMP or CMP (Fig. 5.8a,b).

Salvage pathways

The purine nucleotide IMP can be converted to GMP by one pathway and to AMP via another pathway. There is no direct pathway for the conversion of AMP to GMP or vice versa. However, there are ways in which both AMP and GMP can be converted back to IMP for redistribution to meet cellular needs (Fig. 5.9).

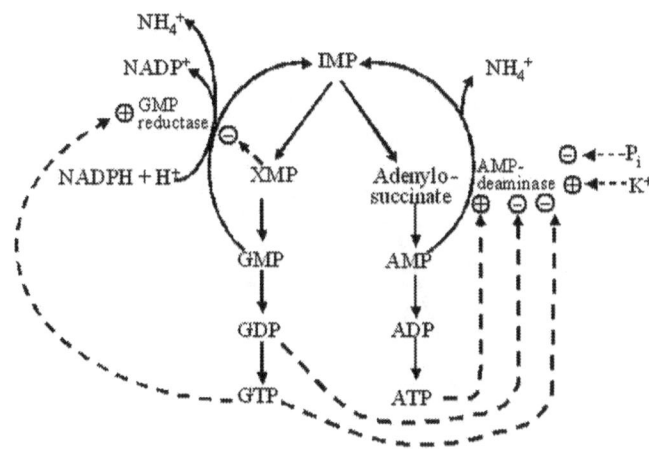

Fig. 5.9: Allosteric regulation in the inter-conversion of purine nucleotides. (-) inactivation; (+) activation; (- - ►) allosteric influence.

Two enzymes carry out these reactions under separate controls.

a) Reductive deamination of GMP to IMP, which is catalysed by GMP reductase, is activated by GTP and inhibited by XMP. XMP, which is a competitive inhibitor of the reductase in humans, has a low K_i (inhibitor constant) of about 0.2 μM. Thus, the concentration of XMP can influence the conversion of GMP to IMP. GTP, as an activator of the reductase, lowers the K_m and increases the V_{max} of the enzyme in respect of GMP.

b) The deamination of AMP, catalysed by AMP deaminase to yield IMP, is activated by ATP or K^+ and inhibited by GDP, GTP or P_i. A plot of V vs. [AMP] in the absence of K^+ is sigmoidal. K^+ or ATP acts as a positive allosteric effector by reducing the K_m of the enzyme for AMP ($K_m app2$) while GTP, GDP or P_i acts as a negative allosteric effector by increasing the K_m for AMP ($K_m app 1$) (Fig. 5.10).

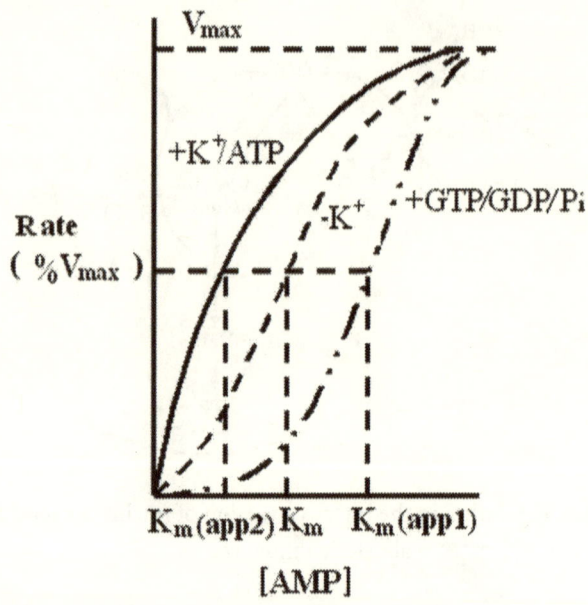

Fig. 5.10: Kinetics of allosteric regulation of AMP deaminase. K^+ or ATP decreases the K_m of enzyme ($K_m$app2) while GTP, GDP or P_i increases the K_m of enzyme ($K_m$app1) for AMP.

Pyrimidine nucleotides

In mammalian cells, regulation of pyrimidine nucleotide synthesis occurs at three steps (1, 3 and 4) in the biosynthetic pathway by allosteric mechanisms (Fig. 5.11). First, at the level of carbamoyl phosphate production from HCO_3^-, NH_4^+ and ATP, catalysed by the cytosolic enzyme carbamoyl phosphate synthetase II, which is allosterically inhibited by UTP (1).   Second, at the level of orate monophosphate (OMP) decarboxylase (3). UMP and to a lesser extent CMP are allosteric inhibitors of OMP decarboxylase. The third regulatory site is the CTP synthetase step (4), which is allosterically inhibited by CTP.

A plot of V vs. [UTP] in the presence of CTP is sigmoidal, i.e. the activity of CTP synthetase is depressed preventing all of UTP from being converted to CTP (Fig. 5.12a).

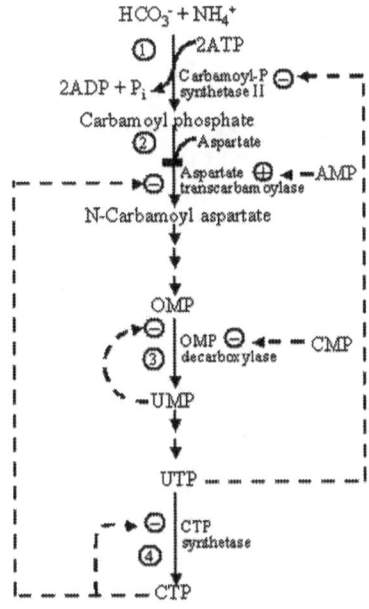

Fig. 5.11: Regulation of pyrimidine nucleotide biosynthesis. (-) inactivation; (+) activation; (-- ▶) allosteric influence.

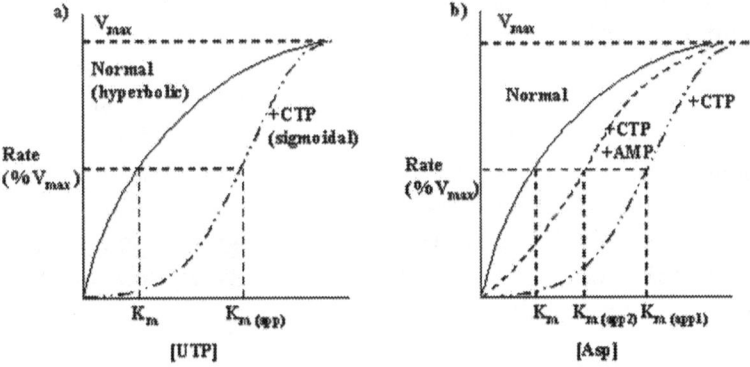

Figs 5.12: Kinetics of allosteric regulation of a) CTP synthetase by CTP and b) aspartate transcarbamoylase (ATCase) by CTP only or CTP and AMP. CTP increases the K_m of CTP synthetase and ATCase for UTP and Asp, respectively. The K_m of ATCase in the presence of both CTP and AMP ($K_m$app 2) is lower than in the presence of CTP only ($K_m$app1), bringing it closer to the normal.

In bacteria (e.g. *E. coli*), the second step (2) in the pathway, catalysed by aspartate transcarbamoylase, is well regulated by CTP and AMP, which are positive and negative allosteric effectors, respectively of the enzyme. This enzyme has six catalytic subunits that bind asp and six regulatory subunits that bind CTP and AMP. Binding of CTP to the regulatory sites causes a conformational change, which is transmitted to the active site leading to inactivation of the enzyme. AMP, when added together with CTP, brings the curve closer to normal by modulating the inhibitory effects of CTP (Fig. 5.12b).

Deoxyribonucleotide synthesis

The synthesis of deoxyribonucleotides is regulated by activators and inhibitors. The binding of these effector molecules regulates the activity and substrate specificity of *E. coli* ribonucleotide reductase involved in the biosynthesis of deoxyribonucleotides (Fig. 5.13).

Adenosine triphosphate activates the reduction of both CDP and UDP. Deoxythymidine triphosphate (dTTP), formed as a result of metabolism of dCDP and dUDP, activates GDP reduction leading to the formation dGDP and accumulation of dGTP, which in turn activates ADP reduction to dADP with consequent formation of dATP. Accumulation of dATP causes the inhibition of the reduction of all substrates. This regulation is further reinforced by the inhibition of the reduction of GDP, UDP and CDP by dGTP and by dTTP inhibition of the reduction of pyrimidine substrates.

There is evidence to suggest that the ribonucleotide reductase step may be the rate-limiting step in deoxyribonucleotide biosynthesis in some eukaryotic cells and, therefore, these allosteric effects may help in regulating deoxyribonucleotide synthesis. A second enzyme in the pathway to dTTP that is subject to allosteric regulation is dCMP deaminase, which is activated by dCTP and inhibited by dTTP. The inhibition of the enzyme by dTTP can be overcome by increasing dCTP concentration.

A plot of V vs. [dCMP] for dCMP deaminase is sigmoidal in the absence of dCTP but hyperbolic in its presence. The latter results in a decrease in the K_m of dCMP deaminase ($K_m$app) for dCMP (Fig. 5.14).

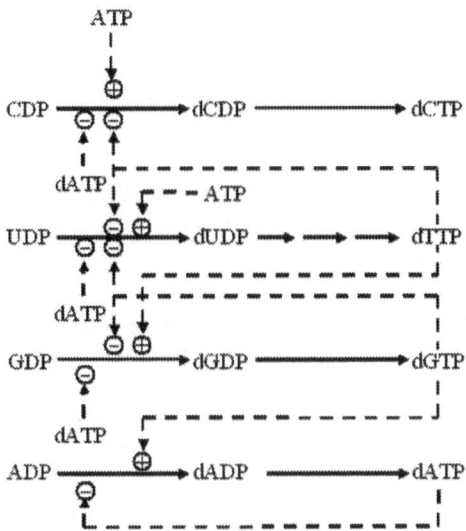

Fig. 5.13: Regulation of deoxyribonucleotide synthesis . (-) inhibition; (+) activation; (--▶) allosteric influence.

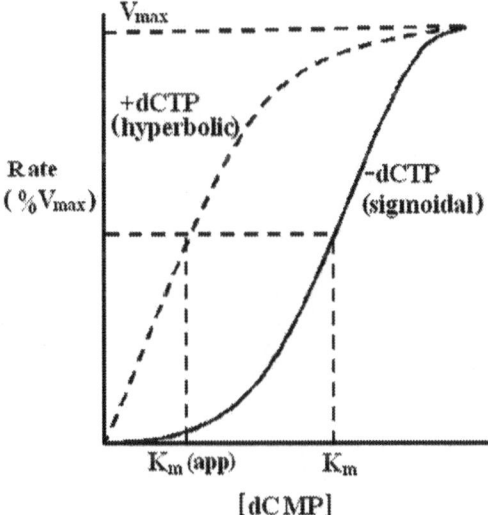

Fig. 5.14: Kinetics of allosteric regulation of dCMP deaminase by dCTP. The K_m of the enzyme for dCMP ($K_m$app) is decreased by dCTP.

Further reading

Bender, D.A. (1985). Amino Acid Metabolism (2d ed.). John Willey and Sons, Toronto

Berg M.J., Tymoczko, L.J., and Stryer, L. (2002). Biochemistry. W.H. Freeman and Company, New York

Christopherson, R.I., and Szabados, F. (1995). Nucleotide Biosynthesis in Mammals. In Agius, L., and Sherratte, H.S.A. (eds.). Channeling in Intermediary Metabolism. London: Portland Pewaa.

Jones, M.E. (1980). Pyrimidine Nucleotide Biosynthesis in Animals: Genes, Enzymes and Regulation of UMP Biosynthesis. *Ann. Rev. Biochem.* 49:253-279

Zubay, L.G. (1998). Biochemistry. Wm. C. Brown Publishers. Boston

PART THREE
INTEGRATION OF METABOLISM

CHAPTER SIX

METABOLIC INTERRELATIONSHIPS

OVERVIEW OF METABOLISM

Metabolic interrelationships are necessitated by the specialized functions of tissues and organs. For instance, unlike other organs, the liver possesses all the enzymes required for the inter-conversion of carbohydrates, lipids and proteins. It is thus able to process the discards of other tissues into much needed fuel molecules for the rest of the body. This cooperation links liver metabolism to that of virtually all non-hepatic tissues. Intricate regulatory mechanisms exist to ensure the equitable distribution of fuel molecules in different physiological states, since fuel reserves are abundant in particular tissues e.g. adipose tissue and liver, but non-existent in others such as the brain and central nervous system.

Tissues and cells exhibit preferences for different fuels based on their peculiar anatomical and biochemical features. For example, the blood-brain barrier prevents the brain from utilizing FAs and the absence of mitochondria in mature RBCs renders them obligate users of glucose. When preferred fuels are in short supply, metabolic inter-conversions become necessary to ensure the survival of such tissues and cells. Tissue differences in expression of genes encoding enzymes such as G-6-Pase, glycerol kinase, β-ketoacyl-CoA transferase impose restrictions on the utilization and fates of certain metabolites, making metabolic integration mandatory.

A steady state exists among pathways as metabolites are continuously added or removed. The steady state can be disturbed by unusual concentrations of substrates and metabolic intermediates, such as in disease states, which tend to direct metabolic processes into unintended pathways.

INTEGRATION OF METABOLISM

Intermediary metabolism is often presented as a set of separate pathways under three sub-headings namely carbohydrate, lipid and protein metabolism. In reality, however, inter-changeability exists among intermediary products of metabolism of these biomolecules and their metabolism is regarded as integrated (Fig. 6.1). Thus, metabolism can be viewed as a single elaborate network of chemical transformations.

Benefits of Integration

The inter-conversion of carbohydrates, proteins and lipids ensures that life does not depend on any particular food type. For instance, even though the brain requires a steady supply of glucose for optimal performance and survival, mammals can survive for prolonged periods without ingesting carbohydrates. Under such circumstances, gluconeogenic amino acids obtained from protein degradation and glycerol from lipolysis are converted to glucose. Another example of the benefits of integration is the ability to convert excess carbohydrates to TAGs, which then supply FAs and glycerol to the body in between meals or during starvation. The FAs serve as fuel molecules and glycerol provides carbons for the synthesis of glucose. Thus, despite variations in diet composition, intermittent food intake and differences in the requirements of organs/tissues, the integrated nature of metabolism enables an organism to obtain energy and specific biomolecules when the need arises.

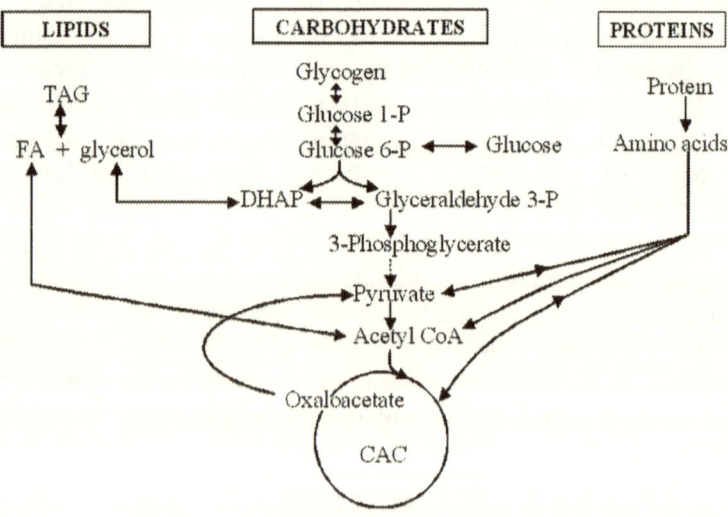

Fig. 6.1 General view of integration of metabolism

Interrelationships between Carbohydrate and Lipid Metabolism

The major points at which pathways in carbohydrate and lipid metabolism intersect are listed below:

a) The catabolism of glucose and other monosaccharides generates pyruvate, which is converted to acetyl-CoA (Fig. 6.1). Acetyl-CoA is of cardinal importance as it is a major crossing point of the metabolism of fats, carbohydrates and proteins. When in excess of what is required for energy generation, acetyl-CoA also serves as an important starting compound for many biosynthetic pathways such as FA and cholesterol synthesis.

b) Glycerol 3-P is needed for the esterification of FAs during the formation of TAGs and it is also used for the synthesis of glycerophospholipids. One source of glycerol 3-P is dihydroxyacetone phosphate (DHAP), an intermediate in the glycolytic pathway.

c) The degradation of TAGs by lipolytic enzymes yields glycerol and FAs. Glycerol is a major precursor for gluconeogenesis.

Interrelationships between Carbohydrate and Protein Metabolism

The major linkages between carbohydrate and protein metabolism are outlined below:

a) During starvation, muscle proteolysis occurs resulting in the release of amino acids, which are subsequently transaminated or deaminated (Fig. 6.1). For example, through transamination alanine, aspartic acid, and glutamic acid yield the pyruvate, oxaloacetate and α-ketoglutarate, respectively. These can then be converted to glucose.

b) Some metabolic intermediates and end products of carbohydrate metabolism serve as precursors for the biosynthesis of non-essential amino acids. For instance, pyruvate and oxaloacetate can undergo transamination to form alanine and aspartic acid, respectively. Serine can be synthesised from the glycolytic intermediate 3-phosphoglycerate.

Interrelationship between Lipid and Protein Metabolism

After a protein-rich meal, some of the excess dietary amino acids are deaminated, and the resulting carbon skeletons converted to acetyl-CoA for the synthesis of TAGs (Fig. 6.1).

Major Junctions in Intermediary Metabolism

The term junction in metabolism refers to any biomolecule that serves as an intersection of two or more metabolic pathways. Numerous minor junctions exist including DHAP and oxaloacetate. These are said to be minor because they can only proceed along two or three metabolic routes. For example, DHAP could participate in glycolysis, gluconeogenesis or the biosynthesis of TAGs. Similarly, oxaloacetate could join the CAC, gluconeogenesis or be used for the synthesis of aspartic acid. In contrast, three of the junctions are described as major. These are G-6-P, pyruvate and acetyl-CoA. Each of these has several fates and precursors, and can be likened to a very busy metabolic intersection (Fig. 6.2). Factors that determine the fates or sources of these biomolecules at any given point in time include cell type, nutritional status, hormonal profile, physical activity and energy requirements. These will be dealt with in detail later on in this chapter.

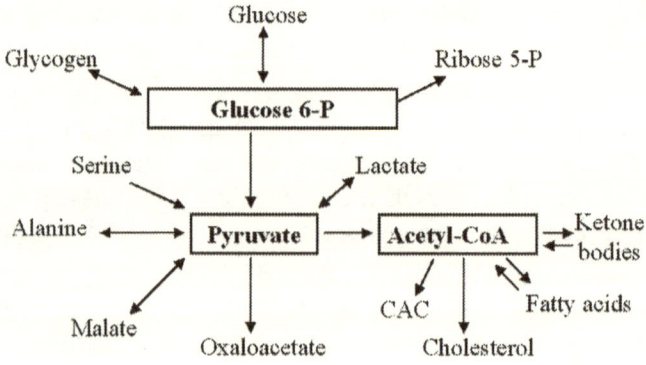

Fig. 6.2 Major junctions in intermediary metabolism

METABOLIC INTEGRATION IN VARIOUS PHYSIOLOG-ICAL STATES

The metabolic pathways that predominate, and the fuel utilisation in a particular tissue at any point in time are determined to a large extent by the nutritional status of the organism (Table 6.1). Insulin, epinephrine and glucagon are the three most prominent hormones that influence cellular metabolism.

THE FED STATE

The fate of dietary fuel molecules in a well-fed state is presented in Fig. 6.3. Insulin exerts its effects on cellular metabolism under such circumstances. Basically, it promotes the utilisation and storage of circulating glucose, amino acids and lipids.

Table 6.1 Fuel availability to various tissues in fed and starved states in man

Tissue	Fed [a]	Starvation Period		
		Early (16-48hr) [b]	Intermediate (2-24days) [c]	Prolonged (>24 days) [d]
Muscle Skeletal	Glucose	Glucose ↓ FFA↑ Ketone bodies↑	Glucose ↓↓ FFA ↑↑ Ketone bodies↑↑	Glucose↓↓ FFA ↑↑ Ketone bodies↑↑↑
Cardiac	Glucose	Glucose ↓ FFA↑ Ketone bodies↑	Glucose↓↓ FFA↑↑ Ketone bodies↑↑	— FFA ↑↑ Ketone bodies↑↑↑
Brain	Glucose	Glucose↓	Glucose↓↓ Ketone bodies↑	Glucose ↓↓ Ketone bodies↑↑↑
Liver	Glucose FFA	— FFA↑	— Glycerol↑ FFA↑↑	— Glycerol↑ FFA↑↑
Adipose tissue	Glucose FFA	— FFA↑	— FFA↑↑	— FFA↑↑

↓ Reduced ↓↓ Markedly reduced ↑ Increased ↑↑ Markedly increased ↑↑↑ Extensively increased

aFuels obtained from nutrients absorbed from the gastrointestinal tract (GIT).

bGlucose obtained from glycogenolysis; free fatty acids (FFA) from lipolysis in adipose tissue and ketone bodies from β-oxidation in muscle and liver.

cGlucose obtained from hepatic gluconeogenesis; FFA from lipolysis in adipose tissue and ketone bodies from β-oxidation in liver.

dGlucose obtained from hepatic and renal gluconeogenesis; glycerol from lipolysis in adipose tissue; ketone bodies from β-oxidation in liver.

Fate of Dietary Carbohydrates

After a meal, digestive enzymes in the gastrointestinal tract (GIT) hydolyze polysaccharides like starch are hydrolysed to monosaccharides, mainly glucose. Their subsequent absorption into the bloodstream causes the blood glucose concentration to rise from the normal 5 mM (80–100 mg/dL) to approximately 8 mM (120–150 mg/dL).

Insulin plays a critical role in the utilisation of glucose after a meal. The pancreas is equipped with glucose sensors and, therefore, in healthy individuals, as blood glucose levels rise, the pancreas responds by secreting greater amounts of insulin. A high insulin to glucagon ratio is obtained and the effects of insulin dominate cellular metabolism. Circulating molecules of insulin bind to specific insulin receptors embedded in the plasma membranes of cells in virtually all tissues.

This triggers a series of reactions that culminate in the activation or inactivation of key enzymes involved in cellular metabolism. Insulin promotes the utilisation of abundant fuel molecules such as glucose by virtually all tissues and storage of the excess. All this is done to bring circulating levels of glucose down to the normal range.

Cellular glucose uptake

Specialized glycoproteins, that serve as glucose transporters (GLUT), are present in the plasma membrane as well as the cytosol of all cells. The rate of glucose uptake can be enhanced considerably by recruiting these cytosolic transporters to the plasma membrane and/or altering their activity. Differences exist in the properties of glucose transporters isolated from different tissues (Table 6.2).

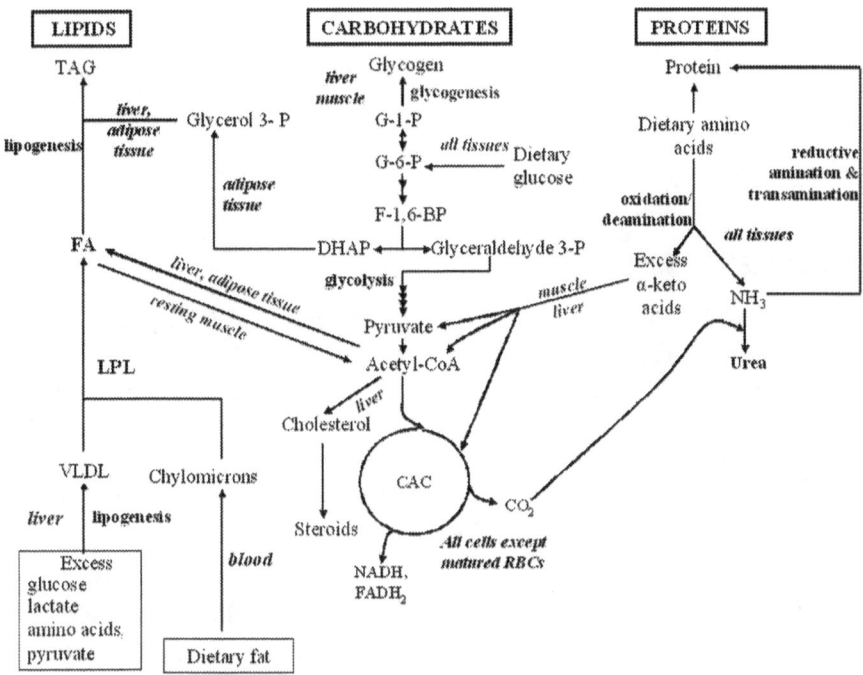

Fig. 6.3: Utilization of fuel molecules in the fed state. Insulin is predominant. LPL = Lipoprotein lipase; VLDL = very low density lipoprotein.

After a meal, glucose is transported from the intestinal epithelial cells into the blood, which is the first tissue to gain access to dietary glucose, and then to the liver. The high K_m for glucose and the insulin-insensitive nature of the transporters in liver ensures that the liver only takes up and stores dietary glucose when circulating levels are very high. This arrangement allows the glucose requirements of the brain, muscle, adipose tissue and other parts of the body to be satisfied.

Since the GLUT4 in adipose tissue and muscle have a low K_m for glucose and are insulin-sensitive, these tissues readily respond to elevations in blood glucose and insulin concentrations by increasing their rates of glucose transport.

Table 6.2 Types of glucose transporters

Type	Affinity	Major sites of expression	Distinctive features
1	High	Brain, RBC, Placenta,	Noninsulin-mediated, constitutive transport
2	Low	Liver, Kidney, Intestine, Pancreatic β-cell	Acts as glucose sensor. Equilibrates intra- and extracellular glucose
3	High	Brain, Placenta	Preferential uptake in hypoglycemia
4	Medium	Muscle, Adipose tissue	Insulin-sensitive

Glucose metabolism

Dietary glucose can be converted to glycogen for storage; acetyl-CoA for FA biosynthesis or energy generation; or ribose and NADPH for biosynthetic processes (Fig. 6.3). Some of the acetyl-CoA is routed to cholesterol biosynthesis. The major consumers of glucose are the brain, red blood cells (RBCs), adipose tissue and muscle. Owing to differences in the functions and composition of tissues and organs there are variations in the fate of glucose in different cell types.

Liver

Insulin stimulates glucokinase, glycogen synthase, PFK-1 and ACC but inhibits glycogen phosphorylase. Therefore, elevated rates of glycolysis, glycogen synthesis, and FA biosynthesis occur when the insulin to glucagon ratio is high. A portion of the glucose that enters the liver is converted to glycogen and the rest of the glucose is metabolised to acetyl-CoA, which can join the citric acid cycle (CAC) or be converted into FA and eventually TAG (Fig. 6.3). The liver is also a major organ for the metabolism of glucose through the PPP for the production of reducing equivalents (NADPH) and ribose 5-P.

Adipose tissue

A major function of adipose tissue is to store excess glucose in the form of TAG. Glycerol 3-P is required for the esterification of FAs, but adipose tissue lacks glycerol kinase, the enzyme responsible for synthesizing glycerol 3-P from glycerol. Consequently, the only source of glycerol 3-P in adipose tissue is the glycolytic intermediate DHAP. This arrangement ensures that adipose tissue is only able to convert glucose to TAGs when DHAP is available. The rates of glycolysis and lipogenesis in adipose tissue are, therefore, closely tied together so that endogenous TAG is only synthesised when carbohydrates are abundant. A fair amount of glucose in adipose tissue is channelled through the PPP to provide NADPH for FA biosynthesis. Insulin stimulates PFK-1 and the PDH complex, leading to a sharp rise in the rate of glycolysis (Fig. 6.3).

Skeletal muscle

Muscle must store glucose as glycogen, a major source of energy during muscular activity. Thus, glucose is either stored as glycogen or is oxidized through glycolysis and the CAC to provide energy. The major actions of insulin in muscle include the stimulation of glycogen synthase, PFK-1 and the PDH complex, and the inhibition of glycogen phosphorylase. As shown in Fig. 6.3, glycogen synthesis, glycolysis and CAC are the dominant metabolic pathways under the influence of insulin.

Other tissues/cells

The brain is a major consumer of glucose and converts it to acetyl-CoA, which is oxidized through the CAC (Fig. 6.3). Mature RBCs obtain energy from glucose via anaerobic glycolysis because they lack mitochondria. Pyruvate is converted to lactate and deposited in the blood. Some of the lactate is taken up by the liver and converted to fat. A portion of the glucose in RBCs is metabolised through the PPP to provide NADPH required for the formation of reduced glutathione, which is essential for the maintenance of cell membrane integrity.

Fate of Dietary Amino Acids

Insulin increases the uptake of amino acids as well as the rate of protein synthesis. Some of the amino acids (glutamine, glutamate, aspartate, asparagine) derived from the dietary proteins are oxidised to serve as a source of energy for intestinal cells. Dietary amino acids are first transported to the liver. Since all tissues require essential amino acids for protein synthesis, the liver allows most dietary amino acids to pass through unless their levels are extremely high, in which case some are used for hepatic protein synthesis. In addition, excess dietary amino acids can be deaminated by the liver and oxidized completely or their carbon skeletons can be used for hepatic lipogenesis. Muscle is capable of oxidizing keto acids obtained from excess dietary amino acids.

Fate of Fat

Dietary

Triacylglycerols make up approximately 90% of the dietary lipids of most humans. The remaining 10% consists of phospholipids, fat-soluble vitamins (A, D, E, and K), cholesterol esters and essential unsaturated fatty acids.

Dietary or exogenous fat enters the blood from the lymphatic system, bypassing the liver, in the form of chylomicrons. In adipose tissue and muscle lipoprotein lipase (LPL), an extracellular enzyme attached to the endothelial cells of the capillaries, hydrolyses exogenous TAGs present in the chylomicrons. While some of the resulting FAs are taken up by adipocytes and esterified with glycerol 3-P to form TAGs for storage, some are oxidized by myocytes for energy.

Endogenous

The TAG synthesized by the liver from excess dietary glucose, amino acids and lactate is packaged in the form of VLDL and released into the bloodstream. In adipose tissue, the fate of VLDL is similar to that of chylomicrons. Cardiac and skeletal muscles take up some of the FAs from VLDL and use them as a major source of fuel.

THE STARVED (FASTED) STATE

Metabolic events in the starved state follow a time course, and can be divided into early, intermediate and prolonged phases (Table 6.1). The dominant circulating hormone in all three phases is glucagon. It increases the intracellular concentration of cAMP, which leads to the activation of cAMP-dependent protein kinase. As a result, glycogen phosphorylase, F-2,6-BPase and PEPCK are stimulated. Consequently, the metabolic pathways that predominate in the liver are glycogenolysis and gluconeogenesis (Figs. 3.10, 3.11, 3.12 and 6.4). Glycogenesis and glycolysis are shut down in order to minimize the utilisation of glucose by the liver and conserve glucose for the brain. Furthermore, by stimulating HSL in adipose tissue, stored TAGs are hydrolysed to release FAs and glycerol. The FAs serve as fuel molecules and glycerol provides carbon skeletons for gluconeogenesis (Fig. 6.4). Ketogenesis and muscle proteolysis prominently feature in the intermediate and prolonged phases of starvation (Fig. 6.4). Details of metabolic events in the three phases are discussed below.

Early Starvation

An average human of 70 kg has sufficient energy stores to survive for 1–3 months without food. During early starvation (16–48 hr), in the absence of dietary carbohydrates, the first priority of metabolism is to provide glucose, the principal fuel for the brain and other obligate users of glucose (Table 6.1).

Because the brain has no fuel reserves of its own it is totally dependent on blood glucose, and to a lesser extent ketone bodies. In the fed state, the average consumption by the brain is 120 g of glucose a day, but this falls to 100 g a day during early starvation. The brain metabolizes it through the glycolytic pathway and the CAC to provide energy. During the first 12–24 hours of starvation, the immediate source of glucose is liver glycogen. Carbohydrate reserves are depleted within a day, after which blood glucose falls to 3–4 mM. Potential precursors of glucose are glycerol from TAG and amino acids from protein breakdown.

Glucagon-induced activation of cAMP-dependent protein kinase results in the phosphorylation and inactivation of several key enzymes in the liver including PFK-2, PK, PDH, glycogen synthase and ACC, while glycogen phosphorylase is activated by the same mechanism. Thus, hepatic glycolysis, glycogenesis and fatty acid synthesis are curtailed while glycogenolysis is activated. Since liver expresses the gene encoding G-6-Pase, G-6-P generated from glycogen degradation is dephosphorylated and released into the blood in an attempt to maintain glucose homeostasis. The inhibition of PFK-2 and PK, which curbs glycolysis, coincides with the promotion of gluconeogenesis. Gluconeogenic amino acids from muscle protein are the chief substrates for both hepatic and renal gluconeogenesis after 48 hours, with glycerol and lactate being used to a lesser extent. The NADH required for the conversion of 1,3-BPglycerate to glyceraldehyde 3-P during gluconeogenesis is supplied during the conversion of malate to oxaloacetate.

The inactivation of ACC leads to low levels of malonyl-CoA in the liver and the consequent lifting of the inhibitory effect of malonyl-CoA on CAT-1. Adipose tissue-derived FAs are then taken up by the liver, transported into the mitochondria and oxidized to yield acetyl-CoA. The enhanced rate of hepatic β-oxidation generates high intracellular concentrations of acetyl-CoA, some of which is oxidized via the CAC to provide sufficient energy to power gluconeogenesis (recall that 6 ATPs or equivalent are required for each molecule of glucose synthesized). At high concentrations, acetyl-CoA allosterically activates PC, which is expressed constitutively in liver. In addition, acetyl-CoA inhibits the PDH complex, ensuring that the rate of gluconeogenesis is elevated, and available pyruvate in the liver is diverted to gluconeogenesis. The cAMP response element binding protein (CREB) can be phosphorylated by cAMP-dependent protein kinase. CREB then binds to an 8-nucleotide DNA sequence termed the cAMP response element, and turns on the transcription of gluconeogenic enzymes. If fasting persists, catabolism is shifted from glucose to FAs and ketone bodies. FAs then become the major fuel and glycerol is made available for the synthesis of glucose. Ketogenic amino acids are metabolized to generate energy and produce ketone bodies, the levels of which are elevated within 12–24 hours.

Events in skeletal muscle differ somewhat from those in the tissues discussed so far. Uptake of glucose by GLUT4 in muscle is diminished because of the low circulating levels of insulin. The principal source of energy is muscle glycogen, which is degraded to yield G-1-P under the influence of epinephrine, whose release is stimulated by low blood glucose concentrations. Unlike liver, muscle lacks G-6-Pase and cannot dephosphorylate the G-6-P obtained from glycogenolysis. This ensures that all the G-6-P derived from muscle glycogen enters the glycolytic pathway for energy generation and none of it is used for the maintenance of blood glucose. The activity of the PDH complex is blocked as a result of epinephrine-induced phosphorylation. Thus, pyruvate obtained from glycolysis cannot enter the CAC, and is converted to either lactate or alanine, both of which are exported to the liver. It is worth mentioning that the properties of the lactate dehydrogenase isoform in muscle favour the formation of lactate while those of the liver isozyme favour the reverse reaction. After approximately 24 hours, when muscle glycogen is depleted, muscle shifts almost entirely to the oxidation of FAs from adipose tissue. Epinephrine also exerts a glycogenolytic effect in liver, which is less pronounced than that in muscle.

Intermediate Starvation

During intermediate starvation (2–24 days), muscle proteolysis provides the bulk of gluconeogenic precursors, and as much as 75 g of muscle is degraded each day. After 7–10 days of starvation, the overall energy expenditure falls significantly, as does the metabolic rate, by up to 30%, leading to decreases in the rates of lipolysis, proteolysis and gluconeogenesis. Gluconeogenesis also occurs in the renal cortex using lactate, glycerol and glutamine (from muscle) as the main precursors. Renal glucose production is stimulated by the high plasma level of epinephrine, and accounts for 20–25% of total after 60 hours of fasting.

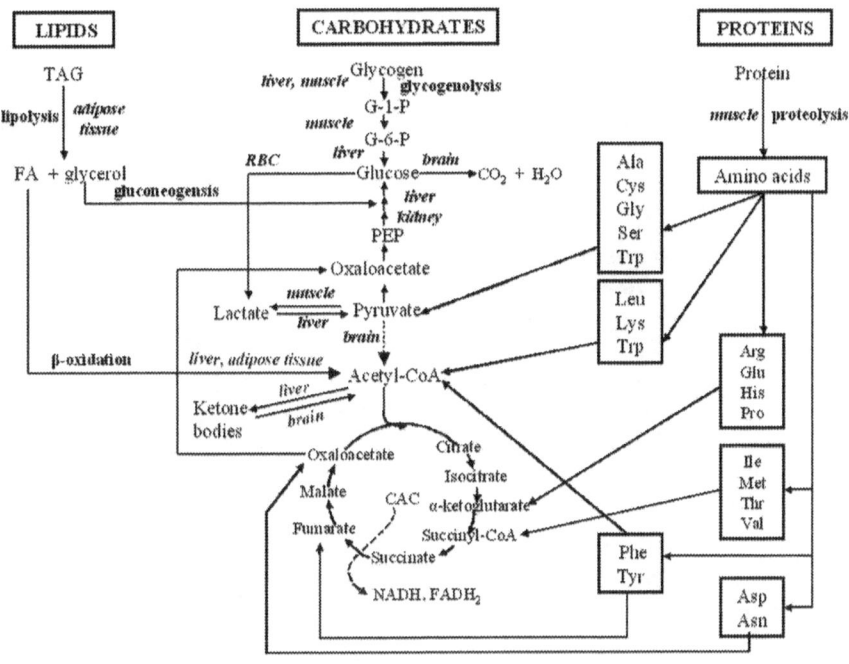

Fig. 6.4: Metabolic interrelationships in the starved state.

The blood concentration of FAs rises from 0.5–0.8 mM to 1.2–6 mM with a corresponding increase in the rate of FA oxidation. In the liver, there is limited CAC activity because most malate is converted to glucose. For this reason, not all the acetyl-CoA generated can be oxidized via the CAC. Instead, ketone bodies are formed and released into the blood. As the plasma concentration of ketone bodies rises, the brain begins to use them to satisfy part of its energy requirements. It is estimated that the brain consumes approximately 50 g of ketone bodies each day in the early part of this phase of starvation and this rises to 100 g as starvation progresses. The increased consumption of ketone bodies by the brain and FAs by most tissues brings about a reduction in the total amount of glucose required by the body. Thus, glucose consumption by the brain drops considerably from 100 g a day in the early part of this phase to 40 g a day later in this phase. As a result, the quantity of gluconeogenic precursors required falls and muscle proteolysis shifts from 75 g per day to 20 g per day. Skeletal muscle oxidizes FAs to spare muscle proteolysis, preventing muscle wasting and thereby prolonging life. The precise mechanisms that trigger muscle proteolysis are not fully understood but growth hormone and cortisol are thought to play major roles.

The general inter-organ relationships between the adipose tissue, liver, muscle and brain in the starved state after depletion of glycogen are shown in Fig. 6.5.

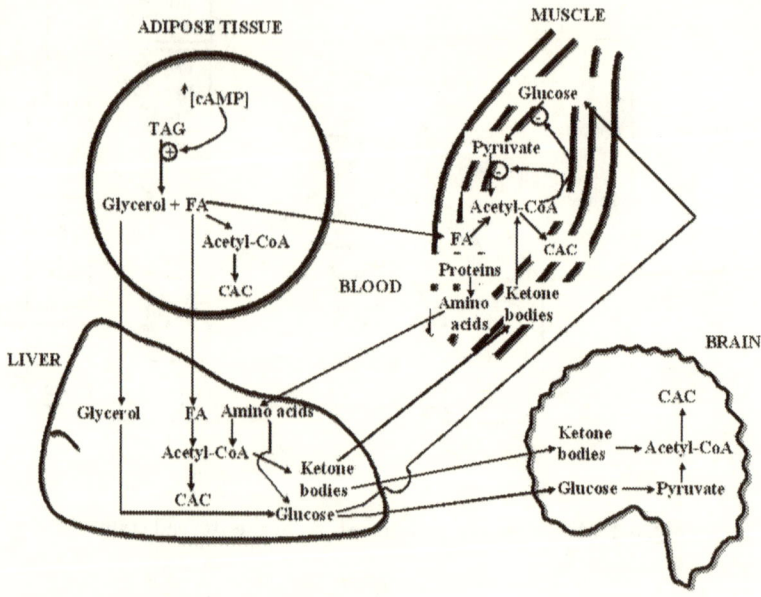

Fig. 6. 5: Inter-organ relationships in the starved state after depletion of glycogen. cAMP stimulates lipolysis while acetyl-CoA inhibits key enzymes in the glycolytic pathway, and the PDH complex

Gluconeogenic precursors

Pyruvate/Lactate

Obligate users of glucose (e.g. brain, RBC and kidney medulla) and muscle continue to use glucose during starvation. The lactate produced by the RBCs and muscle is carried to the liver via the blood and converted back to glucose in the inter-organ cycle called **Cori cycle** (Fig. 6.6). The energy required for the synthesis of glucose is obtained from β-oxidation of FAs. Thus, FA oxidation contributes to glucose homeostasis even though FA cannot be converted to glucose directly.

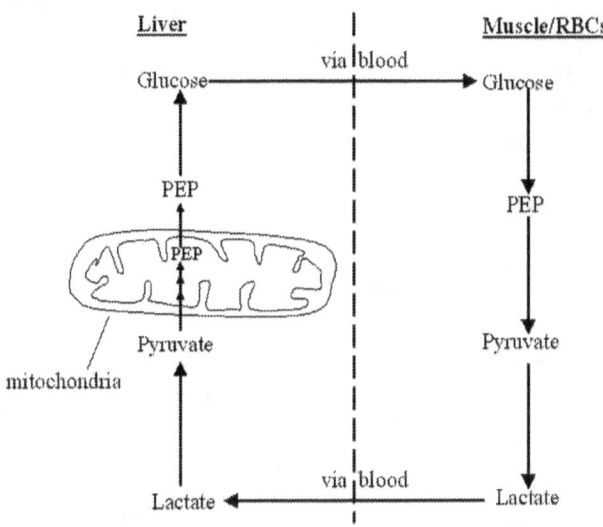

Fig. 6.6: The Cori cycle: It operates in intermediate starvation or in active muscles. Glucose synthesised in liver from lactate through gluconeogenesis by use of energy from β-oxidation of FAs is transported to peripheral tissues like muscle, kidney medulla and brain to be utilised, and the lactate produced through glycolysis is transported back to the liver for further conversion to glucose.

Amino acids

During this period of starvation 100–110 g of glucose are used per day. Provision of 100 g of glucose requires the breakdown of 175 g of protein. Muscle at this stage is in a state of net proteolysis, which accounts for about 40% of the glucose produced by hepatic gluconeogenesis. The release of amino acids produces intermediates important for gluconeogenesis (e.g. oxaloacetate and citrate). Most branched-chain amino acids and aspartic acid are metabolised in the muscle.

Ammonia produced by the oxidative degradation of amino acids from muscle proteolysis is carried from the muscle to the liver through ala by means of the inter-organ cycle called the **glucose-alanine cycle** (Fig. 6.7). This together with the Cori cycle accounts for 40–45% of the daily glucose needs of the body through gluconeogenesis. The NH_3 is first converted to glutamate in the presence of α-ketoglutarate catalysed by glutamate dehydrogenase. The

NH_3 of glutamate is transferred to pyruvate, produced from metabolism of glucose in the muscle via glycolysis, by alanine aminotransferase to form ala. Thus, ala carries NH_3 from the muscle to the liver via the blood.

In the liver, alanine aminotransferase catalyses the conversion of ala and α-ketoglutarate to pyruvate and glutamate. The pyruvate is converted to glucose, which is transferred from the liver to the brain. The glutamate undergoes deamination to produce α-ketoglutarate and NH_3, which is converted to urea and excreted.

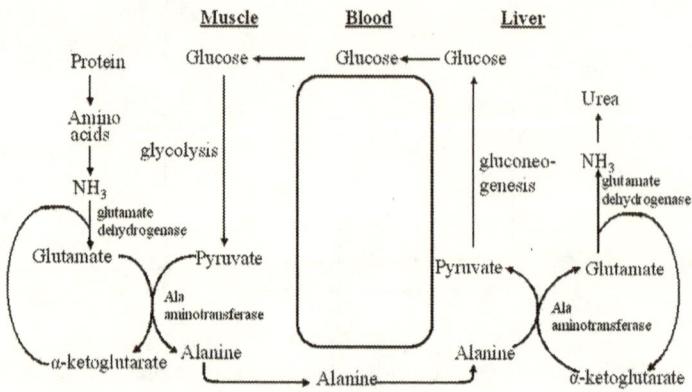

Fig. 6.7: The glucose-alanine cycle. It carries amino groups of glutamate via alanine from skeletal muscle to liver to be converted to urea and provides working muscle with glucose made by liver from the breakdown of alanine

Glycerol

Glycerol continues to be produced and converted to glucose at a faster rate at this stage. It is phosphorylated by glycerol kinase in the cytosol to form glycerol 3-P, which is converted to DHAP either by 1) a cytosolic glycerol 3-P dehydrogenase or 2) a mitochondrial membrane glycerol 3-P dehydrogenase complex that requires quinone (Fig. 6.8). DHAP can then be converted to glucose by the reversal of glycolysis.

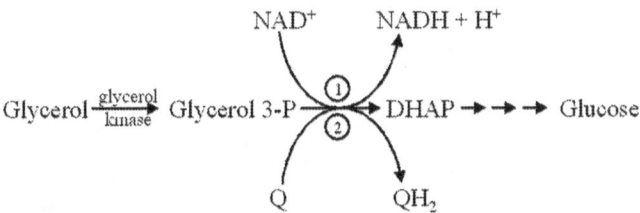

Fig. 6.8: Hepatic gluconeogenesis using glycerol as substrate. Conversion of glycerol 3-P to DHAP is catalysed by (1) cytosolic or (2) mitochondrial membrane glycerol 3-P dehydrogenase. Q = Quinone; QH_2 = Dihydroquinone.

Ketone bodies

Ketone bodies are formed by the liver as a result of the high rate of production of acetyl-CoA, far in excess of what the CAC can metabolise. They are taken up by the brain and muscle and metabolised for the provision of energy (see Chapter 4), thus sparing the use of glucose. It is estimated that ketone bodies provide 50% of the energy needs of the muscle and spare the use of branched chain amino acids produced by muscle proteolysis. Ketone bodies cross the blood-brain barrier at certain threshold blood concentrations, enter the brain and serve as fuel molecules to the brain. In the small intestines they spare the use of glutamine whose metabolism by the kidneys produces bicarbonate (HCO_3^-) for buffering blood. In addition glutamine metabolism generates NH_3 for buffering of urine in ketoacidosis and glucose for utilisation by brain in late starvation.

The overall effect of the use of ketone bodies is that there is less demand for glucose and less muscle proteolysis. Thus, much of muscle protein is reserved mainly for prolonged starvation.

Prolonged Starvation

Glucose

The brain utilises glucose during prolonged starvation (>24 days into starvation), but at a markedly reduced rate (Table 6.1). The presence of GLUT3 transporters (Table 6.2) enables the brain to take up glucose even when blood glucose levels fall between 2.2–5.0 mM. Other tissues and cells, namely RBCs, renal medulla, eye lens and cornea also utilise some glucose. During this phase, renal gluconeogenesis assumes greater importance and may produce up to

50% of circulating glucose. If blood glucose concentration falls below 2.2 mM, the brain becomes dysfunctional leading to coma and subsequently death.

Fatty acids/ketone bodies

Ketone bodies continue to be oxidised by the brain and muscle during prolonged starvation while FA oxidation occurs primarily in muscle and liver (Table 6.1). The blood-brain barrier is impermeable to FAs and thus prevents the uptake of FAs by the brain. In general, the duration of starvation depends to a large extent on the amount of TAG reserves present. Obese individuals tend to withstand longer periods of starvation because it takes longer to exhaust their supplies of FAs from fat reserves.

In very prolonged starvation, virtually every tissue utilises either FAs or ketone bodies or both. Energy expenditure is greatly reduced to prolong life. For instance, the resting metabolic rate decreases to approximately 70% of normal after 30 days of starvation. Nevertheless, starvation cannot continue indefinitely. Eventually, when adipose tissue disappears, muscle protein becomes the final source of energy. The muscles waste away as a result of the degradation of essential proteins and consequently death becomes imminent.

THE RUNNING ATHLETE

The dominant circulating hormone at the start of a race in a healthy and well-fed athlete is epinephrine. The sources of energy utilized and metabolic patterns that pertain depend on the intensity and duration of the race. Detailed actions of epinephrine have been discussed in Chapters 3 and 4.

Sprinter

During short duration, high power output exercise such as in (world class) sprinting, muscle operates anaerobically because the fast twitch muscle fibres involved have few mitochondria, and the demand for ATP far exceeds the rate of oxygen delivery to these muscle cells. Since oxygen is the final acceptor of electrons in the ETC, the production of sufficient ATP by this route is not feasible. The initial source of energy available to the runner is from pre-existing ATP and phosphocreatine until muscle glycogen stores are mobilised under the influence of epinephrine, and anaerobic glycolysis becomes the most important source of ATP in muscle (Fig. 6.9).

The phosphorylation and activation of glycogen phosphorylase by high levels of epinephrine lead to the stimulation of glycogenolysis, resulting in the formation of G-6-P (Fig. 6.9). The release of Ca^{2+} during initial muscle contraction also contributes to the initiation of glycogenolysis through allosteric activation of phosphorylase kinase. Muscle lacks the enzyme G-6-Pase and since phosphorylated glucose cannot be exported, all the G-6-P generated by the degradation of muscle glycogen is used for the production of energy to support muscular activity. Glucose uptake by muscle is also enhanced. The interaction between epinephrine and its receptors in skeletal muscle causes an increase in the rate of glycolysis. This occurs via cAMP-mediated phosphorylation and activation of PFK-2, which leads to a rise in the intracellular concentration of F-2,6-BP, a positive allosteric effector of PFK-1. Epinephrine-induced phosphorylation and inhibition of the PDH complex together with the anaerobic state of muscle cause pyruvate to be converted to lactate (Fig. 6.9) with the production of 3 moles of ATP per mole of G-6-P. It is estimated that muscle glycogen stores are sufficient to sustain approximately 80 seconds of such high intensity exercise.

The effects exerted by epinephrine and glucagon in the liver result in an uninterrupted supply of blood glucose to the muscle. Together, these hormones promote hepatic glycogenolysis and gluconeogenesis. The liver isoform of PFK-2 is inhibited when phosphorylated, curtailing hepatic glycolysis. Unlike muscle, liver is endowed with G-6-Pase and is thus able to dephosphorylate the G-6-P obtained from glycogenolysis and gluconeogenesis. Glucose is released into the bloodstream, making it available for the muscle. Lactate produced by the muscle during this period is transported to the liver, converted to glucose via gluconeogenesis. The glucose produced is then transported to the anaerobic muscle to be utilised in the Cori cycle (Fig. 6.9). The muscle is unable to utilize FAs under these circumstances.

Marathon Runner

During long duration and relatively moderate physical activity, such as a marathon, slow twitch muscle fibres characterized by abundant mitochondria and high levels of myoglobin operate aerobically. Epinephrine interacts with skeletal muscle, adipose tissue and liver to ensure that adequate amounts of fuel molecules are released to sustain the exercise (Fig. 6.9). The stimulatory effects of epinephrine and glucagon on hepatic glycogenolysis and gluconeogenesis provide the muscle with glucose. Muscle glycogenolysis, which is enhanced through the activation of glycogen phosphorylase, is gradual and can last for two hours. Glucose uptake by muscle is also enhanced, as is the rate of glycolysis (Fig. 6.9). Under such conditions, the CAC in muscle operates fully because muscle is aerobic.

Aerobic skeletal muscles largely obtain energy from FAs. It is estimated that the oxidation of FAs can supply as much as 60% of total energy for at least four hours of long distance running. Adipose tissue TAG stores are mobilised as a result of high glucagon and epinephrine levels to provide FAs, which can be metabolised through β-oxidation in the liver and muscle to acetyl-CoA. Since the CAC in both tissues is fully operational under these conditions, acetyl-CoA is completely oxidized to provide energy. Glycerol obtained from the hydrolysis of adipose tissue TAGs is taken up by the liver and serves as a gluconeogenic precursor. AMP-activated protein kinase (AMPK) is emerging as a key regulator of metabolic processes in muscle during exercise. Experimental evidence suggests that it enhances fat oxidation and ACC phosphorylation during exercise.

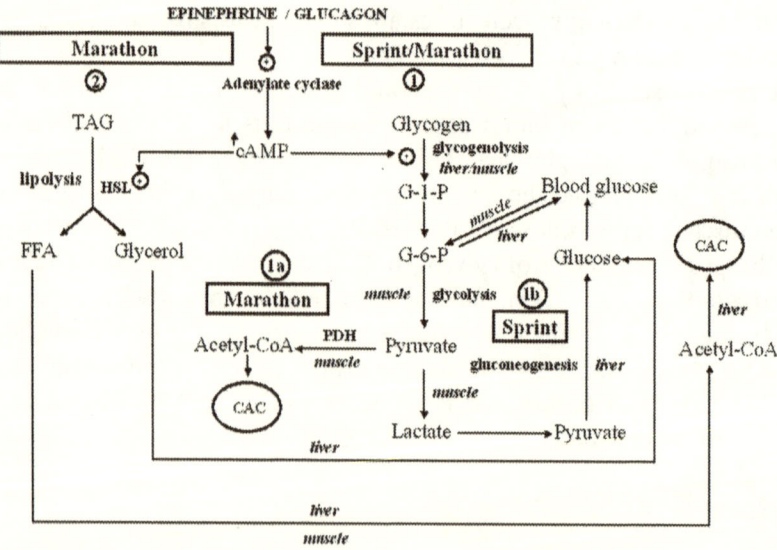

Fig. 6.9 Metabolic interrelationships in a running athlete: during intense physical activity, e.g. a sprint race (1b) or during a moderate activity, e.g. a marathon race (1a & 2). HSL = Hormone sensitive lipase; PDH = Pyruvate Dehydrogenase; CAC = Citric acid cycle.

THE DIABETIC STATE

Diabetes mellitus DM is a disease state characterized by hyperglycaemia due to either decreased production of insulin or decreased sensitivity to insulin. There are two main types. Insulin-dependent diabetes mellitus

(IDDM) also known as juvenile-onset or Type 1 diabetes, which is associated with autoimmune destruction of the β-pancreatic cells is characterized by insulin deficiency. Non-insulin dependent diabetes mellitus (NIDDM) or Type 2 diabetes often manifests after age 40 and is, therefore, termed maturity-onset diabetes. The hallmark feature of the latter is insulin resistance in the presence of normal or elevated plasma insulin levels. It has been postulated that Type 2 diabetes results from defects in insulin binding or post-receptor insulin signalling mechanisms.

IDDM-Metabolic Interrelationships

In IDDM, elevated postprandial glucose levels fail to trigger the release of appropriate quantities of insulin from the pancreas. Since insulin normally promotes the cellular uptake, utilization and storage of glucose, the deficiency in insulin impairs entry of glucose into cells as well as its utilization, causing hyperglycaemia. In addition, elevated levels of glucagon stimulate glycogen phosphorylase and F-1,6-BPase but inhibit glycogen synthase, PFK-1 and PK. The overall effect of these actions is the mobilization of stored glucose and increased production of glucose by the liver, exacerbating the hyperglycaemic condition (Fig. 6.10).

Glucagon also elevates the rate of lipolysis in adipose tissue, an action that results in the release of large amounts of FAs into the bloodstream, and increased FA uptake by the liver and tissues other than the adipose tissue. The major fuels for hepatic gluconeogenesis are glycerol from adipose tissue and amino acids from the degradation of muscle protein. The liver, through β-oxidation, converts some of the FA to acetyl-CoA. Low levels of intracellular oxaloacetate, due to its use in gluconeogenesis, make it difficult to oxidize all the acetyl-CoA via the CAC. Excess acetyl-CoA is converted to ketone bodies (Fig. 6.10). Ketoacidosis occurs when the rate of utilisation of ketone bodies by extrahepatic tissues is not as rapid as the rate of synthesis by the liver. The liver is unable to process all the circulating FAs in this manner and consequently, a sizeable proportion is esterified by the liver to TAGs, packaged as VLDL and released into the blood. LPL, the enzyme responsible for clearing VLDL from the blood, is stimulated by insulin in normal individuals. However, LPL remains in an unstimulated state in IDDM subjects. Thus, the rate at which VLDL is released into the blood by the liver far exceeds the rate of clearance by LPL, resulting in hypertriglyceridemia. Hyperchylomicronemia is also associated with this condition since LPL is the enzyme that removes chylomicrons from the blood.

The situation in all tissues is that of starvation despite the abundance of dietary carbohydrates, lipids and proteins. The metabolic pathways that

predominate are glycogenolysis, gluconeogenesis, FA oxidation and ketogenesis in liver, lipolysis in adipose tissue and muscle proteolysis (Fig. 6.10).

NIDDM-Metabolic Interrelationships

The aberrations in carbohydrate metabolism are similar to those in IDDM,
but differences exist in lipid metabolism. In subjects with NIDDM, circulating
levels of FFA and TAG are elevated, and there is excessive deposition of TAG in
muscle and tissues other than adipose tissue, but ketoacidosis is rare. Several
factors contribute to the hypertriglyceridemia. In healthy individuals, insulin
exerts a potent suppressive effect on HSL, the principal regulator of FFA release
from adipose tissue, and stimulatory effects on FFA esterification. However, in
NIDDM subjects there is reduced insulin-mediated suppression of HSL resulting in increased lipolysis in adipose tissue. As alluded to earlier, esterification of
FAs in adipose tissue depends on the availability of DHAP, the precursor for
glycerol 3-P. Since glucose uptake by GLUT4 is diminished and glycolysis is curtailed, little glycerol 3-P is available for the esterification and re-esterification of
FAs. Consequently, postprandial FA disposal is impaired. The combined effects
of accelerated lipolysis and impaired esterification is hypertriglyceridemia and
an increase in the uptake of FAs by liver and muscle. This abnormal metabolic
situation is further compounded by the fact that muscle exhibits a reduced
capacity to oxidize FAs for reasons that are not entirely clear.

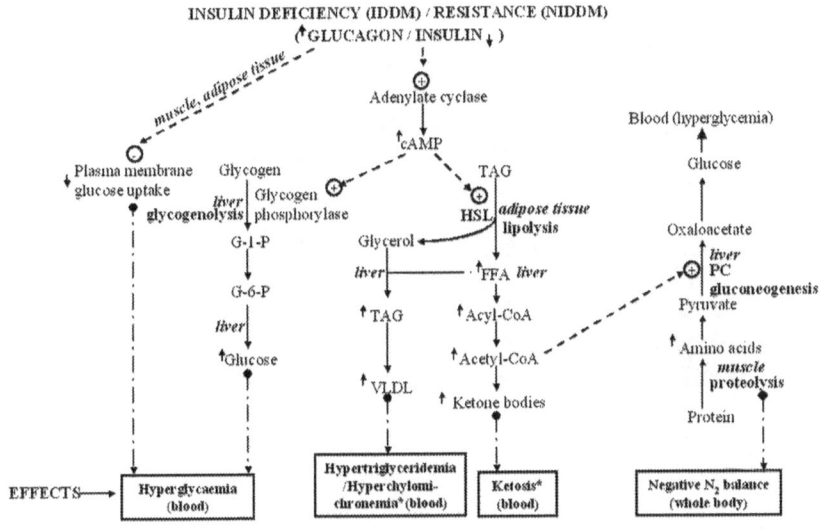

Fig. 6.10: Metabolic interrelationships in the diabetic state. *Absent in NIDDM; PDH = Pyruvate dehydrogenase; PC = Pyruvate carboxylase; HSL = Hormone sensitive lipase; ACC = Acetyl-CoA carboxylase; PFK-1 = Phosphofructokinase-1; VLDL = Very low density lipoproteins; ↓(decreased level); ↑(increased level).

Hepatic esterification is favoured over β-oxidation and the high level of insulin leads to high hepatic expression of sterol regulatory element binding protein-1c (STREB-1c), a member of a family of regulated transcription factors, which enhances the transcription of genes for ACC and FA synthase. The resultant chronic stimulation of *de novo* lipogenesis greatly contributes to an elevated rate of VLDL production and hypertriglyceridemia. Hepatic malonyl-CoA concentration increases and esterification of incoming FAs is favoured over oxidation. Newly synthesized TAGs are either stored in the cytosol or released into the blood as VLDL. However, beyond a threshold concentration of long chain acyl-CoAs, the inhibitory effect of malonyl-CoA on CAT 1 is counterbalanced such that enhanced FA oxidation and esterification occur simultaneously. It is believed that reduced VLDL clearance may also contribute to the high plasma VLDL levels because insulin activation of adipose tissue LPL activity is delayed in certain categories of Type 2 diabetics. A high rate of VLDL production also raises the plasma concentration of intestinally derived chylomicrons because of the competition between chylomicrons and VLDL for removal from the plasma by LPL leading to hyperchylomicronemia.

Further reading

Bollen, M., Keppens, S., and Stalmans, W. (1998). Specific Features of Glycogen Metabolism in the Liver. *Biochem. J.* 336:19-31

Coven, D.L., Hu, X., Cong, L., Bergeron, R., Shulman, G.I., Hardie, D.G., and Young, L. H. (2003). Physiological Role of AMP—Activated Protein Kinase in the Heart: Graded Activation During Exercise. *Am. J. Physiol. Endocrinol. Metab.* 285:E629-E636

Czech, M. P., and Corvera, S. (1999). Signaling Mechanisms that Regulate Glucose Transport. *J. Biol. Chem.* 274:1865-1868

Holloszy, J.O., and Kohrt, W.M. (1996). Regulation of Carbohydrate and Fat Metabolism During and After Exercise. *Ann. Rev. Nutr.*16:121-138

Jose-Cunilleras, E., Hinchcliff, K.W., Sams, R.A., Devor, S.T., and Linderman, J.K. (2002). Glycemic Index of a Meal Fed Before Exercise Alters Substrate Use and Glucose Flux in Exercising Horses. *J. Appl.Physiol.* 92: 117-128

Nordlie, R.C., Foster, J.D., and Lange, A.J. (1999). Regulation of Glucose Production by the Liver. *Ann. Rev. Nutr.* 19:379-406

Seaquist, E.R., Damberg, G.S., Tkac, I., and Gruetter, R. (2001). The Effect of Insulin on *In Vivo* Cerebral Glucose Concentrations and Rates of Glucose Transport/Metabolism in Humans. *Diabetes.* 50:2203-2209

Spriet, L.L., and Watt, M.J., (2003). Regulatory Mechanisms in the Interactions between Carbohydrate and Lipid Oxidation during Exercise. *Acta Physiol. Scand.* 178:443-452

Standaert, M.L., Ortmeyer, H.K., Sajan, M.P., Bandyopadhyay, G., Hansen, B.C., and Farese, R.V. (2002). Skeletal Muscle Insulin Resistance in Obesity-Associated Type 2 Diabetes in Monkeys Is Linked to a Defect in Insulin Activation of Protein Kinase C—/zeta/lambda/iota. *Diabetes.* 51:2936-2943

ABOUT THE AUTHORS

Naa A. Adamafio obtained a B.Sc. (Hons) in Biochemistry from the University of Ghana in 1977 and a Ph.D. in Biochemistry at Monash University, Melbourne, Australia under the Australian Commonwealth Fellowship Plan. She was a postdoctoral fellow and a research investigator at the University of Michigan, Ann Arbor, USA under the American Diabetes Association fellowship/grant. A seasoned endocrine biochemist, she teaches courses on Integration and Control of Metabolism, Biochemistry of Hormones and Signal Transduction. She is at present a Senior Lecturer in Biochemistry and the Head of the Biochemistry Department of the University of Ghana.

Laud K. N. Okine graduated with B.Sc. (Hons) Biochemistry from the University of Ghana in 1975 and obtained his PhD in Biochemical Toxicology at the University of Surrey, UK under the British Commonwealth Fellowship Plan. He was a Visiting Research Fellow at the National Cancer Institute, National Institutes of Health, USA, under the John Fogarty International Fellowship Award. He has taught for many years on the subjects of Intermediary Metabolism, Metabolic Regulation, Xenobiotic Metabolism and Clinical Biochemistry. He is a Senior lecturer at the Department of Biochemistry, University of Ghana, and also the Director of a reputable Research and Development establishment in Plant Medicine in Ghana.

Jonathan P. Adjimani obtained his B.Sc. (Hons) Biochemistry at the Kwame Nkrumah University of Science and Technology in 1979, M.Sc. Biological Science from Brock University, Canada and Ph.D. Biochemistry from the Utah State University, USA. He was a visiting professor at Brigham Young University, Provo, Utah. He was a holder of many academic awards including the Research Vice-President's Fellowship (Utah State University) and the Lady Davis Postdoctoral Fellowship (Hebrew University of Jerusalem) Awards. He has taught for many years on the subjects of Intermediary Metabolism, Enzymology and Molecular Biology. At present, he is a Senior Lecturer at the Department of Biochemistry, University of Ghana.

ILLUSTRATION CREDITS

Chapter 2: Fig. 2.7b Adapted from Moran, L. A. *et al.* (1994), Biochemistry, Neil Patterson Publishers/Prentice Hall Inc, Englewood Cliffs, NJ, USA, p. 12.38; Figs. 2.8-2.11 From Matthews, C. K. and Van Holde, K. E. (1996), Biochemistry, The Benjamin/Cummings Publishing Co. Inc, Menlo Park, CA, USA, pp. 973, 974, 977, 988.

Chapter 3: Figs. 3.4, 3.8, 3.10-3.14 Adapted from Devlin, T. M. (1992), Textbook of Biochemistry with Clinical Correlations, Wiley-Liss Inc, New York, NY, USA, pp. 252, 322, 323, 335, 346, 347, 352-354, 356; Fig. 3.9 Adapted from Nelson, D. L. and Cox, M. M. (2000), Lehninger Principles of Biochemistry, Worth Publishers, New York, NY, USA, p. 587.

Chapter 4: Figs. 4.1-4.3 Adapted from Moran, L. A. *et al.* (1994), Biochemistry, Neil Patterson Publishers/Prentice Hall Inc., Englewood Cliffs, NJ, USA, pp. 20.23, 20.46.

Chapter 5: Figs. 5.3, 5.4 Adapted from Moran, L. A. *et al.* (1994), Biochemistry, Neil Patterson Publishers/Prentice Hall Inc, Englewood Cliffs, NJ, USA, pp. 21.6 and 21.7; Fig. 5.5 Adapted from Matthews, C. K. and Van Holde, K. E. (1996), Biochemistry, The Benjamin/Cummings Publishing Co Inc, Menlo Park, CA, USA, p. 766; Figs. 5.6 and 5.7 Adapted from Nelson, D. L. and Cox, M. M. (2000), Lehninger Principles of Biochemistry, Worth Publishers, New York, NY, USA, pp. 840, 853; Fig. 5.9 Adapted from Devlin, T. M. (1992), Textbook of Biochemistry with Clinical Correlations, Wiley-Liss Inc, New York, NY, USA, p. 544; Fig. 5.13 Adapted from Zubay, G. L. *et al.* (1995), Principles of Biochemistry, Wm. C. Brown Publishers/Wm. C. Brown Communications Inc, IA, USA, p. 558.

Chapter 6: Fig. 6.7 Adapted from Zubay, G. L. *et al.* (1995), Principles of Biochemistry, Wm. C. Brown Publishers/Wm. C. Brown Communications Inc, IA, USA, p. 521.

ABBREVIATIONS USED IN TEXT

ACC	Acetyl-CoA carboxylase
Ala	Alanine
AMP, ADP, ATP	Adenosine 5' -mono-,di-, triphosphate
AMPK	AMP-activated protein kinase
Asp	Aspartic acid (aspartate)
ATase	Adenylyltransferase
ATCase	Aspartate transcarbamoylase
CAC	Citric acid cycle
cAMP	3', 5'-Cyclic AMP
CAP	Catabolite activator protein
CAT	Carnitine acyl transferase
CMP, CDP, CTP	Cytidine 5'-mono-di-, triphosphate
CoA(SH)	Coenzyme A
CREB	cAMP response element bindung protein
CRP	cAMP receptor protein
dADP, dATP	Deoxyadenosine 5'-di-, triphosphate
DAG	Diacylglycerol
dCMP, dCDP, dCTP	Deoxyctidine 5'-mono-, di-, triphosphate
dGDP, dGTP	Deoxyguanosine 5'-di-, triphosphate
DHAP	Dihydroxyacetone phosphate
DM	Diabetes mellitus
DNA	Deoxyribonucleic acid
dTTP	Deoxythymidine 5'- triphosphate
dUDP	Deoxyuridine 5'-diphosphate
E	Enzyme
eEF(s)	Eukaryotic elongation factor(s)
eIF(s)	Eukaryotic initiation factor(s)
ER	Endoplasmic reticulum
ES	Enzyme-substrate complex
ETC	Electron transport chain
FA(s)	Fatty acid(s)

FAD, FADH	Flavine adenine dinucleotide and its reduced form
FFA(s)	Free fatty acid(s)
F-1,6-BP(ase)	Fructose 1,6-bisphosphate (ase)
F-2,6-BP(ase)	Fructose 2,6-bisphosphate(ase)
F-6-P	Fructose 6-phosphate
DG	Free energy change
G-1-P	Glucose 1-phosphate
G-6-P (ase)	Glucose 6-phosphate (ase)
G-6-PD	Glucose 6-phosphate dehydrogenase
GIT	Gastrointestinal tract
GLUT	Glucose transporter
Gly	Glycine
GMP, GDP, GTP	Guanosine 5'-mono-,di-, triphosphate
His	Histidine
HMG-CoA	3-Hydroxy-3-methylglutaryl-CoA
HSL	Hormone sensitive lipase
I-1a	Inhibitor 1a
IDDM	Insulin-dependent diabetes mellitus
Ile	Isoleucine
IMP	Inosine 5-monophosphate
IP3	Inosine 1, 4, 5-triphosphate
IPTG	Isopropylthiogalactoside
IRS(s)	Insulin receptor substrate(s)
k_1, k_{-1}	Rate constant for forward and reverse reactions of an equilibrium reaction
k_{cat}	Turnover number
K_{eq}	Equilibrum constant
K_i	Inhibitor constant
LPL	Lipoprotein lipase
PC	Pyruvate carboxylase
PD (cAMP)	3', 5'-Cyclic AMP phosphodiestrase
PDH	Pyruvate dehydrogenase
PEP	Phosphoenolpyruvate

PEPCK	Phosphoenolpyruvate carboxykinase	RBC (s)	Red blood cell (s)
		RNA	Ribonucleic acid
PFK-1, PFK-2	Phosphofructokinase-1 & 2	S	Substrate
P_i	Inorganic phosphate	Ser	Serine
PIP_2	Phosphatidyl inositol 4,5-bisphosphate	SR	Sarcoplasmic reticulum
		SS	Steady state
PK	Pyruvate kinase	STREB-1c	Sterol regulatory element binding protein 1c
PKA	Protein kinase A		
PKC	Protein kinase C	T	Tense/tight
		TAG(s)	Triacylglycerol(s)
P13K	Phosphatidylinositol 3-kinase	Thr	Threonine
		Trp	Tryptophan
PLC	Phospholipase C	UMP, UDP, UTP	Uridine 5'-mono-, di-, triphosphate
PP1	Phophoprotein phosphatase 1		
		V, V_{max}	Velocity, maximum velocity
PPP	Pentose phosphate pathway	VLDL	Very low-density lipoprotein
PRPP	5-Phosphoribosyl-1-pyrophosphate		
		XMP	Xanthine 5'-monophosphate
R	Relaxed		

INDEX

978-0-595-34067-5
0-595-34067-9

www.ingramcontent.com/pod-product-compliance
Lightning Source LLC
Chambersburg PA
CBHW030745180526
45163CB00003B/924